"In this book, Dr. Yaq
cent discoveries. He is abl
niques using a very simple language and make those more accessible to the general audience."

"The author talks to the reader and keeps the reader wanting
more and more. It was hard to avoid flipping fast to the next page
and see what was next."

Prof. D. F. de Mello, Washington, D.C.

Exoplanets

and

Alien Solar Systems

Tahir Yaqoob, Ph.D.

A publication of New Earth Labs™ (Education and Outreach),
P. O. Box 5672, Baltimore, MD 21210.
http://www.newearthlabs.com.
Further information and updates on *Exoplanets and Alien Solar Systems*
can be found at http://exoplanets.co.

Also by the same author: *What Can I Do to Help My Child with Math
When I Don't Know Any Myself?* Available in eBook and paperback
formats. http://helpyourchildwithmath.com.

ISBN: 978-0-9741689-2-0

Cataloging data:

Exoplanets and Alien Solar Systems.
Tahir Yaqoob.
Includes index.
Keywords: 1. Exoplanets; 2. Extrasolar planets; 3. Astronomy;
4. Astrophysics; 5. Physics; 6. Space Science; 7. Planetary Science.

Contents

Chapter 1

Stepping Out

Connecting with a Conversation Through History

Whatever you were doing before reading this, take your mind away from it, and imagine that you are outside, barefoot on a pebbly beach, looking out at the vibrant blue ocean in front of you. It's slightly chilly. Feel the cool wind running against your face. Breathe in and smell the rawness of the Earth around you. Clouds move out of the way and the warmth of the Sun caresses your face. Hear the relentless but soothing rhythm of the wavy water. Feel the hard stone under your feet. Your eyes feast on the view of the blue abyss in front of you. You are being nurtured from head to toe. But you are standing on a planet that is hurtling through an unforgiving vacuum at an unimaginable speed of over $60,000$ miles per hour (about $100,000$ kilometers per hour). Yet you stand peacefully as one, with this colossal ball of rock and water that protects you and gives you life. Up in the sky rages a hydrogen furnace of incredible dimensions, that has faithfully served your planet for five billion years, so that it may give you life.

In the modern trappings of existence it is so easy to forget that you live on a planet. Just beyond home is an environment that is so lethal that it's hard to believe the picture described above could ex-

1

ist. Cosmic rays that could devastate your atomic make up. Electric and magnetic fields that could instantly put your lights out. X-rays and other radiation that could fry you in an instant. Temperatures so cold that you would shatter into a billion fragments. A vacuum that would suck the life out of you. All of this is not very far away. Yet you can stand on your planet, look around at the beauty that cocoons you, and feel like you are the most fortunate being in the Universe.

In this book we will connect with a conversation that has permeated humanity throughout history. A conversation that is the manifestation of a collective consciousness that goes back to a time before we even knew what a planet was, before we knew how to make fire, all the way back to whenever self-awareness arose. A very long chain of developments spanning billions of years (starting from the first cells) led finally, in the 1960s, to one species on Earth leaving the planet. Now, since the 1990s humanity has entered a new phase in which we can now take a peek at other planets outside our solar system. When we talk about planets beyond our solar system, remember that we are continuing an ancient endeavor to understand ourselves. There is something that is at once fascinating, mind-boggling, and even a little bit sad about this. Imagine going back in time half a million years or so and meeting with your ancestors. Even if you could communicate with them, how would you even begin to explain to your ancestors what a planet is? You could not do it. The collective conceptual machinery of the human race had not advanced sufficiently for your exposition to have any meaning. A stupendous number of advances in conceptualization and knowledge that spanned thousands of generations had to be made first. If you downloaded your knowledge of the planets, gravity, atomic structure, stellar evolution, and the like onto your ancestors you would be branded as a deviant nutter.

Our collective consciousness and knowledge, something that does not reside in any single human, has ended up in a place that could not have been predicted by our ancestors half a million years ago (for example). Now look the other way, into the future. Not a thousand years, not ten thousand years, not a hundred thousand years. What about another half a million years, this time in the future? Would it not be too presumptuous to think that we can predict where our collective knowledge and consciousness is going, and where it will be at that time? Is it not honest to admit that we may understand things then that we have no idea how to understand now? Don't you think that there are things that exist *now*, about which we have no knowledge and about which we cannot even conceive a question? Our ancestors half a million years ago (I am using this number as an arbitrary but sufficiently ancient point in time) could never have conceived the question, "Are there planets beyond our solar system?" Yet those planets existed nevertheless.

At the risk of sounding too melodramatic, as you read this book and a wide variety of others like it, immerse yourself in the idea that you are partaking in something that is as ancient and as grand as the origin of life itself. It is not just an idea. It is a very real phenomenon, as new knowledge, awareness, and understanding become assimilated into the minds of individual members of the species like you.

Exoplanets Reveal Themselves

Back in the late 1990s, reports of observations of planets outside our solar system were becoming more and more frequent. One of the methods used to find and study exoplanets involves monitoring the occultation of the host star by the planet, as it moves in its orbit.

3

Although I had dabbled in the occultation technique in a different context that had nothing to do with planets, I remained skeptical of the findings. [1] The skepticism was quite widespread amongst scientists. The other methods used to find exoplanets seemed too tentative and subject to uncertainties. Personally, when I was a kid interested in space science and the search for other planets that could support life, I did not expect that planets outside our solar system would be discovered in my lifetime. Somehow, I was still holding on to that when the exoplanets *were* being discovered.

Then, sometime in the early 2000s, everything changed. The evidence just became too overwhelming and hard to refute. Many people had to undergo a mental paradigm shift. This was real. It was really happening. For so long, so many had wanted this, and now it was happening in my lifetime. The story is still unraveling in real time. My own research is not directly in the field of exoplanets, [2] and I am only just one of many messengers. My hat goes off to the thousands of people over several decades who have been directly responsible for making the new science of exoplanets real, and to the millions of people like you who are interested enough to have enabled all of this to happen. It is hard to imagine that any exploration of our solar system and others beyond it would have happened if there were no interest amongst the general public.

I believe that developments at the frontiers of human knowledge should be widely accessible to all, so the main text in this book deliberately avoids jargon, equations, and unnecessary technical detail. I would like this book to be a nonstressful (I hesitate to say relaxing), casual read. All concepts at the bottom of our attempts to understand physics and math come from the human mind, and are usually very simple. They only become complicated when they are expressed and applied in certain contexts. (Whether the concepts underlying the physics and/or math correspond to reality is only verified or refuted after they are generated in the mind.) For the reader with a greater background in the relevant technical material, there are plenty of annotated notes throughout the text that can serve as a basis for pursuing further, more in-depth knowledge.

The new science of exoplanets is in a very nascent phase and is obviously advancing extremely rapidly. A book like this can quickly get out of date. However, I have written as much of it as possible in a way that will make the information durable. In particular, I have tried to show you what the key open issues are, and how to interpret (and place into context) new results as they come in.

A note about the use of the phrase "solar system." Strictly speaking, "solar" is a term associated with our Sun, whereas "stellar" is a term associated with any star in general. However, the term "solar system" is in use in the popular and specialist literature as a label that is not exclusive to our Sun and is also used for systems associated with other stars. Therefore, I will also use the term "solar system" to refer to a system of planets around a star other than the Sun, although it is not, strictly speaking, correct to do so.

Jupiter's Compelling Attraction

Back in the late 1980s, some of us were pondering deep questions such as whether or not we like CDs, and what was going to happen to our LP collection. Some time in that period somebody at a party once said to me, "Did you know that the gravitational attraction between a newborn baby and the planet Jupiter is greater than the gravitational attraction between the mother and her baby?" (I didn't say it was a good party!) As you may already be aware, gravity plays such a critical role in the formation, dynamics, and evolution of all celestial bodies in the Universe that we will constantly be running into it. In particular, we will very frequently come across the fact that planets that are as massive as Jupiter can have a powerful gravitational influence on other planets in a multiplanet system (Jupiter is about 318 times more massive than Earth), even at separation distances that are larger than the Earth-Sun distance. In our solar system the formation of the Earth would have proceeded differently for the two cases that Jupiter had or had not yet formed. Even now, Jupiter can bump off asteroids out of their regular orbits. Jupiter also induces gravitational stresses in some of its moons (which in turn can cause heating and volcanic activity).

As it happens, the geeky question that was posed in the preceding paragraph actually serves as an excellent illustration of how powerful a large and massive planet like Jupiter can affect the formation, dynamics, and fate of a planetary system. In fact we will see that the answer to the mother/baby question actually depends on *how* the mother holds the baby.

The force of gravity between two masses is directly proportional to the product of the masses but decreases as the distance between the centers of the two masses increases (with the same mathematical dependence as the brightness of a light as you move further away from it). Let's take two extreme cases, one in which the mother and baby are engaged in a close hug, and the other in which the mother holds the baby out at arm's length. Using the current Earth-Jupiter distance, I have made some estimates of the ratio of the gravitational force between mother and baby, and the force between baby and Jupiter. [3] In the close-hug situation, the gravitational attraction between mother and baby is stronger than that between baby and Jupiter, but when the baby is held out at arm's length, the converse is true and Jupiter takes over. The apparent coincidences are such that, with a 70 kg (about 160 pound) mother, the baby-Jupiter attraction is about two-thirds of the baby-mother attraction in a close hug, but increases to about 17 times the baby-mother attraction when the baby is held at arm's length.

The results do not depend on the mass of the baby: it "drops out." The results also do not depend on where on planet Earth mother and baby are located: the largest margin of error is tiny, no more than 0.004%. There is some dependence on the mass of the mother. The Jupiter-Earth distance varies, but even during Jupiter's closest approach, any mother with a mass greater than about 50 kg (110 pounds) will dominate over Jupiter in a close hug. Moreover, Jupiter will *always* dominate when the baby is held at arm's length, unless the mother is more massive than about $1,200$ kg ($2,640$ pounds), which is of course unlikely. We should bear in mind that the *absolute* forces we are talking about are tiny: for a seven-pound (3.2 kg) baby, the attraction between Jupiter and the baby is equivalent to about one ten-millionth of a pound in weight. Despite the tiny absolute value, the exercise *does* tell us something very important about planet formation.

You see, the current theory of planet formation says that rocks and boulders came together under their own gravity to form progressively larger and larger lumps of matter that eventually became planets. Now suppose that the mother and baby are two boulders (or other space debris) that are trying to come together in the process of forming Earth. Well, the above calculations have just shown that if the rocks are more than about a meter (about 3 feet) apart, then Jupiter is going to dominate and try to pull them away. They will have a hard time coming together. So if Earth formed after Jupiter, the conditions for formation would have been constrained by the presence of Jupiter. I will discuss planet formation, and the problems associated with understanding it, in much more detail later in the book. However, the estimates here should give you a feel for why Jupiter-like planets can influence planetary formation and solar system dynamics so heavily.

Chapter 2

A Sketch of the New Science

Chapter Overview

Before we start talking about exoplanets we should really talk about what a planet is. However, since the discovery of objects comparable in size and mass to Pluto, and since the demotion of Pluto, the official definition (as expounded by the *International Astronomical Union*, or IAU) is actually tediously boring and full of "sticky issues," so I have put a detailed discussion in appendix A. You should read it when you feel up to it. For the moment, all you need to know is that for exoplanets there are a sufficiently large number of "issues" that the only thing that is certain is a mass upper limit for something to be called a planet. This upper limit is about 12 to 13 Jupiter masses (depending on the details of the calculation), and corresponds to the point at which an object starts to ignite nuclear burning of deuterium ("heavy hydrogen," or hydrogen atoms that have two neutrons instead of one). Objects that are more massive than this upper limit are called brown dwarfs. (At about 80 Jupiter masses, nuclear burning of regular hydrogen can begin and the object could potentially become a "regular" star.) There is no formal lower limit on an object to be called a planet (but see appendix A). In the end, as far as I am concerned, it doesn't really matter whether we call something a planet or not, because the ultimate goal is to understand every kind of object that is found, in its own right. Our classification of these objects is bound to evolve as our understanding of them improves and there are bound to be objects in the "grey" areas of classification schemes that are based on phenomenology and incomplete knowledge.

In the remainder of this chapter we will take a quick look (a sketch) of where we are at this remarkable juncture in human history. We will take a whirlwind tour of our own solar system to provide a quick recap and to get you up to speed on our state of knowledge. This will be useful for putting the new exoplanet discoveries in context. Obviously it is going to be a *very* quick tour, given the size and scope of this book. However, it should be sufficient for gaining an appreciation of the new discoveries concerning exoplanets. There are plenty of notes in the text to guide you on where to find additional information and resources in addition to the selected list of websites in appendix B. Then we will cover some basics about the population of exoplanets, some of the terminology, and what sort of things we would like to know about exoplanets and the systems they reside in. However, I will need to refer frequently to the distances to the planets and to star-planet distances, so first I will establish a means of expressing relative distances in a way that will give you a "feel" for those distances.

Near and Far

When scientists talk about distances in our solar system and distances between planets and stars in other solar systems, they will most often use the term "AU," or astronomical unit. The AU is based on the Earth-Sun distance but the latter is variable because the orbit is "slightly" elliptical and not circular. Therefore, the *International Astronomical Union* (IAU) has come up with a working definition that gives a single number, whereby 1 AU is equal to a specific distance accurate to more decimal places than you might care for. For our purpose it is sufficient to say that 1 AU is approximately 149.6 million kilometers (or approximately 93 million miles). The actual derivation of the exact number is stupendously boring so I will spare you of those details. To get a feel for the AU, consider that at a speed of 20, 000 miles per hour (achievable with current space travel), it would take about 194 days to cover 1 AU. As another kind of benchmark, consider that the "circumference" of the Earth's orbit is about 6.3 AU, and that it obviously takes the Earth a year to cover this! (The Earth's velocity around the Sun is therefore about 66, 705 miles per hour, or 107, 350 kilometers per hour.) In the next section you will get more of a feel for the AU in terms of other "benchmarks" in the solar system.

When talking about distances between stars, there is a problem with the AU because it is so much smaller than the distances between stars. Several different units are used by astronomers for interstellar distances and the one that is probably most appropriate for our purpose is the light year. The light year is often confused with being a unit of time, but it is a *distance*. That distance is approximately 5, 878 billion miles (9, 640 billion kilometers), which is how far light would travel in one year, at its speed of approximately 670 million miles per hour (or about 1.1 billion kilometers per hour). Sometimes a distance may be given in light minutes or even light seconds. In each case the meaning is analogous to the meaning of a light year. A useful benchmark is that the distance between the Earth and the Sun is about 8 light minutes. In other words, 1 AU is about 8 light minutes, which means it takes 8 minutes for the light from the Sun to reach Earth. The Earth-Moon distance is about 1.3 light seconds (approximately 239,000 miles, or about 384,000 kilometers).

Solar System Overview

Vast amounts of knowledge have been gained about our solar system since I was a kid, more than I imagined to be possible in that time. Just as exciting is the fact that important mysteries remain, and there is still much to be learned.

As we shall see later, exoplanets exhibit a larger variety of physical properties than the planets in our solar system and include types of planets that are not represented in the solar system. We can think of the planets in the solar system in terms of two qualitatively different groups. The four inner planets (known as terrestrial, or rocky planets) are dominated by material that is on average 4 to 5.5 times denser than water. This is indicative of significant quantities of substances containing heavy elements such as silicon, iron, nickel, and others. Silicates are particularly abundant in rocky planets. The distance between the Sun and the terrestrial planets lies in the range of approximately 0.4 to 1.5 AU, going from Mercury, Venus, Earth, out to Mars. All four terrestrial planets are about the same size as each other. Venus is within a few percent of the Earth's size, whilst Mercury and Mars are about 40% and 50% of the Earth's size respectively. Since all of the terrestrial planets have about the same density, this means that Mercury and Mars are much less massive than Earth (Mercury and Mars have about 5.5% and 10% of the mass of Earth respectively). Venus has about 80% of the mass of the Earth. [1]

Planets in the second group are characterized by the absence of a solid surface. Instead, they have substantial envelopes of hydrogen and helium, the two most abundant elements in the Universe.

12

These four outermost planets lie at a distance of about 5 AU to 30 AU from the Sun, Jupiter being the closest to the Sun, followed by Saturn, Uranus, and Neptune, in that order. All of these outer four planets are larger than the inner terrestrial planets, with Jupiter and Saturn being the largest (about 11 and 9 times the diameter of Earth respectively). Uranus and Neptune are both about 4 times the size of Earth. It is thought that the outer four "giant" planets probably have compact rocky cores, but it is currently hard to verify or investigate this directly. All four outer planets are characteristically less dense than the terrestrial planets by factors of 2.4 to 7.9 (approximately). As we shall see for exoplanets, the average density is a key factor in guessing what the chemical composition of a planet might be because no other information might be available. Saturn is actually the only planet that is less dense than water (by about 30%). The other three outer planets vary in density, with the highest density being about 65% higher than water (for Neptune). This can be compared with the terrestrial planets, which have densities that are approximately 300% to 450% higher than water. For the solar system planets (as opposed to exoplanets), a lot of information about their composition is available in addition to density, and it has been known for a long time that the low densities of the outer planets are due to a heavy weighting towards lighter elements, in particular hydrogen and helium. Whereas Saturn and Jupiter have hydrogen and helium envelopes that dominate in both mass and volume, in Uranus and Neptune hydrogen and helium dominate in volume but not mass. Uranus and Neptune harbor methane and various ices in addition to the hydrogen and helium envelopes. Thus, although all four outer planets are extremely cold, Jupiter and Saturn are commonly known as gas giants and Uranus and Neptune are commonly known as ice giants. Jupiter and Saturn have an effective temperature of about -150 and -180

13

degrees Celsius respectively, and Uranus and Neptune are about 40 degrees Celsius colder than Saturn. (In Fahrenheit units, that is −240 and −290 degrees for Jupiter and Saturn respectively, and Uranus and Neptune are about 70 degrees cooler than Saturn.) In terms of mass, the four outer planets again fall into two groups, with Jupiter and Saturn being much more massive than Uranus and Neptune. The enormous mass of Jupiter (about 318 Earth masses) and its wide-ranging influence on solar system dynamics and evolution has already been mentioned. Saturn has about 95 times the mass of Earth, whilst Uranus and Neptune have about 14.5 and 17.1 times the mass of Earth. When we come to talking about exoplanets we will see that Earth, Jupiter, and Neptune have become benchmarks, in terms of mass, size, and sometimes composition, for qualitatively characterizing exoplanets.

To put things into perspective, Jupiter, despite its extremely large mass, still only has about 0.1% of the mass of the Sun, but it is about 10% of the size of the Sun. The density of Jupiter is similar to that of the Sun, which is also composed predominantly of hydrogen and helium. A new NASA mission, *Juno*, was launched on the 5th of August 2011 to study key physical properties of Jupiter (e.g., its atmosphere below the cloud cover and its gravitational and magnetic fields). The expected arrival time of *Juno* is July 2016 and the mission is expected to yield some of the most detailed properties of Jupiter yet.

Jupiter and Saturn are well-known for entertaining several large moons with some very interesting properties, including, in some cases the possibility of supporting life. It is beyond the scope of this book to go into detail about the satellites of the planets in the solar system. However, what is relevant for the purpose of this book is that our solar system demonstrates that satellites with significant masses and sizes exist, and that the possibility of detecting exomoons (satellites around exoplanets) is promising. To this end, it is worth bearing in mind that the largest satellite in the solar system (Ganymede around Jupiter) is larger than Mercury and Pluto, and it is about 40% of the size of Earth.

In between Mars (at about 1.5 AU from the Sun) and Jupiter (which moves between about 4.2 to 6.5 AU from the Sun), is the asteroid belt, consisting of a large number of lumps of matter, most of them irregular in shape. Asteroids occur in a range of sizes and there are likely to be more than the ones that are known (more than half a million). However, their total mass can be dynamically determined and adds up to less than the mass of our Moon (which itself is only 81 times less than the mass of the Earth). The composition of the asteroids varies. Some are dominated by carbonaceous compounds, some by silicates (i.e., by a rocky composition), whilst others are metallic (iron/nickel). Asteroids are particularly interesting because they represent some of the most ancient components of the solar system, unaffected by weathering. Some of the larger asteroids are spherical. In particular, the largest asteroid, Ceres, is spherical and it has qualified for classification as one of the five "official" dwarf planets of the solar system (but refer to a full discussion in appendix A). The origin of the asteroids is thought to be the result of the destruction of a former planet by differential gravitational stress (tidal forces) from Jupiter, and/or by a catastrophic collision.

Tiny moons have been found around some asteroids, the first of these discovered is known as Dactyl, just a mile wide, orbiting an asteroid known as Ida, which is itself only 19 miles wide.

The NASA mission, *Dawn*, was launched in September 2007 to study asteroids and achieved an orbit around the asteroid Vesta in July 2011. In 2012 it is then to go on to study the largest object in the asteroid belt, Ceres, a dwarf planet. The mission is already

15

revealing interesting findings and is expected to return the most detailed information on the physical properties of individual asteroids to date.

Some asteroids can be in unstable orbits due to the influence of Jupiter's powerful gravitational field, and may be knocked out of the asteroid belt, potentially ending in a catastrophic collision course with some other body in the solar system. As for finding the equivalent of an asteroid belt in an alien solar system, it is unlikely with current capabilities because most asteroids are so small and are thus comparatively weak reflectors of light. However, it is conceivable that a "smoking gun" could be detected, in terms of, for example, an impacted exoplanet.

Whether a rocky body in the solar system has an atmosphere or not is important in determining the temperature and climate properties of that body, which in turn affect its abilities to support life. The factors that determine whether a planet, moon, or other object is able to hold on to an atmosphere, and that determine the composition of the atmosphere if it has one, are complex. I will discuss these in more detail at relevant points throughout the book. For the moment we should appreciate the fact that even though Earth and Venus are so similar in size, mass, and distance from the Sun, they have very different climates. This fact alone tells us that the issues are complex. Venus has a thick atmosphere of carbon dioxide, and a brutal greenhouse effect, supporting surface temperatures of greater than 450 degrees Celsius (nearly 900 degrees Fahrenheit). Atmospheric composition can change with time, as well as with biological activity. It is possible that the terrestrial planets in the solar system had large hydrogen and helium envelopes in the past that have since been lost due to evaporation and escape from the host planet's gravitational field (something that atmospheres composed of the lighter elements are more prone to). Moreover, on

Earth, biological activity has changed the composition of the atmosphere over long periods of time: the oxygen was not a significant constituent of Earth's atmosphere before biological processing.

Mercury and the Earth's Moon are two bodies in the solar system that are well-known for having no atmosphere. Until recently, it was thought that Pluto did not have an atmosphere either, but a carbon monoxide atmosphere associated with Pluto has now been discovered. [2] The spatial extent of the atmosphere is huge, reaching out to approximately three times Pluto's size. There is even evidence of variability in the composition of the atmosphere on a timescale of ten years, and of a comet-like morphology of the atmosphere, but these things still need to be confirmed. It is important to remember that Pluto is very different to the terrestrial planets and is smaller than the Earth and our Moon in both mass and size. Pluto has only 0.2% of the mass of Earth, and is about a third of the size of Earth. Its low average density (approximately twice that of water) means that it cannot be dominated by a rocky composition. Pluto is a cold, icy world, being so far from the Sun. The core may be rocky, under layers of water ice, although methane and nitrogen ices cover the surface. Pluto's highly eccentric orbit can take it as far as about 49 AU from the Sun, but it can also take it closer to the Sun than Neptune, approaching a distance of just under 30 AU from the Sun. Pluto has been very difficult to study, but the very first mission to visit the dwarf planet, *New Horizons*, will reach its destination in 2015.

Beyond Pluto lies a murky region called the "Kuiper Belt" that is essentially a region containing space junk, or rocky/icy debris. A lot of these objects would now qualify to be classified as "dwarf planets." However, the *International Astronomical Union* (IAU) has infinite wisdom and officially recognizes only 5 dwarf planets in our solar system (Ceres the asteroid, Pluto, Eris, Haumea and Makemake). [3] The reader is referred to appendix A for a detailed discussion of the challenges involved in planet and dwarf planet classification. The other objects in the Kuiper Belt are known as Kuiper Belt Objects, or KBOs (don't fall off you chair in amazement!). They are also known as Trans-Neptunian Objects, or TNOs. More than 1, 300 KBOs have been identified, but it is extremely difficult to measure their sizes because they are so far and faint. The *New Horizons* mission will go on to study the Kuiper Belt (which is thought to extend from about 30 AU to 50 AU) after it has left Pluto.

In reality Pluto is really an inner KBO and it was the discovery of other sizable KBOs that then began the path to the eventual reconsideration of whether Pluto is a planet or not. The large KBO known as Quaoar was discovered in 2002 and has a diameter of about 800 miles (1, 287 kilometers), which is more than half of the diameter of Pluto. [4] The dwarf planets Eris and Haumea are comparable in size to Pluto, and the dwarf planet Makemake is *thought* to be comparable in size to Pluto. The Kuiper Belt is also a reservoir for short-period comets (generally speaking, comets that take less than about 200 years to complete an orbit around the Sun).

The Kuiper Belt is thought to have a definite edge at about 50 AU, an inference that has been made from a measured drop-off in KBOs from surveys. [5] Such surveys also yield estimates of the total number of KBOs of 70, 000 or so. Certainly, objects in the Kuiper Belt that have very eccentric orbits can venture outside of the formal outer edge of the Belt, but there appears to be a definite cutoff in the population of KBOs beyond 50 AU. This brings us to another "exotic" member of our solar system, *Sedna*. This is an icy body, discovered in 2004, that merits the title of the most distant object ever seen in our solar system. Sedna's size is very difficult to measure but a diameter of about 800 to 1, 100 miles (1, 287 to

1, 770 kilometers) has been estimated (i.e., about the same size as Quaoar). [6] Sedna is not a KBO for some very important reasons. For one thing, Sedna never gets closer to the Sun than about 76 AU and therefore never enters the Kuiper Belt. In fact, the Sun is not enclosed in Sedna's orbit so it cannot even be defined as a dwarf planet (the discovery paper called it a planetoid). Sedna's orbit is so eccentric that the planetoid travels as far as 1, 000 AU or so from the Sun. What is extremely puzzling is the mechanism that could have placed Sedna into such an orbit. If Sedna had been gravitationally scattered by Neptune, some part of the orbit should again approach Neptune, but Sedna's orbit passes nowhere near Neptune. The Sedna discovery paper states, "Such an orbit is unexpected in our current understanding of the solar system." Since the discovery, searches for similar objects have been unsuccessful. [7] A satisfactory explanation of Sedna's orbit has not been found. Speculative ideas include scattering by an undiscovered planet, a "close" encounter with another star, and formation of the solar system in a star cluster. However, more information is required in order to make any robust deductions. The implications of the very existence of Sedna are of course profound, because it represents a dynamical relic of the early solar system and very likely holds clues pertaining to the true scenario of the formation of the solar system and probably planet formation in general.

Sedna is regarded as a member of the inner Oort cloud. The Oort cloud is a *postulated* region beyond the Kuiper Belt that was invoked (in 1950, well before the discovery of Sedna) as a reservoir of long-period comets. Since the Oort cloud is not an observed entity, only a hypothesized region, estimates of the locations of its inner and outer edges vary wildly in the literature. If Sedna is a member of the Oort cloud then its distance from the Sun of approximately 76 AU to 1, 000 AU obviously locates the innermost estimate of the inner edge of the Oort cloud. As for the outer edge, you can find up to $200,000$ AU for its value in the literature. This upper end actually corresponds to more than 3 light years, so at such vast distances objects are only loosely bound to the host star and are subject to potential influence from other nearby stars.

Finally, before leaving the solar system, there is one more "exotic" object I will mention. You may have heard of a giant planet in the outer solar system (in the Oort cloud) called Tyche. This is something that has not actually been observed, but rather it has been *deduced* to exist by means of the analysis and interpretation of the orbital elements of long-period comets. A paper published in 2010 that presented the case for Tyche, suggested that the mass of the hidden object is between 1 and 4 times the mass of Jupiter. [8] The paper is very careful not to call the object a planet, and simply refers to the object as a "Jovian-mass solar companion." There are assumptions that go into the theory and analysis, and it is too early yet for the methodology and interpretation to have been decisively scrutinized by other researchers.

As far as exoplanets are concerned, I will summarize what can be taken away from all of the above discussion about the solar system. Firstly, there is still so much that we do not know about our own solar system, especially about its early history and formation, that solar system research needs to be pursued as vigorously as exoplanet research. In particular, the Kuiper Belt and the vast region beyond it, hold important clues that should help to eventually understand exoplanet systems as well as our own solar system. Current capabilities do not yet allow direct observation of the equivalent of KBOs or Oort cloud constituents in exoplanet systems, which makes it even more compelling to try to better understand our own solar system.

One other thing that will be important when we look at exoplanets and alien solar systems is the inclination angle and eccentricity of a planet's orbit around its host star. The inclination angle is an indicator of the alignment of the plane of the planetary orbit with respect to the sight line to the observer. The eccentricity is a measure of how "squashed" the shape of the orbit is compared to a circle. Yet another important quantity is the angle between the spin axis of a host star and the orbital plane of an exoplanet associated with that host star. If this angle is not 90 degrees, the orbit is skewed with respect to the spin of the star. We shall see that there are examples of exoplanets whose orbits are highly skewed relative to the spin axis of the host star, and examples of exoplanet orbits that have high eccentricities (even for exoplanets that are not very distant from their host stars). Skewness and eccentricity are potential clues about a planet's formation and/or collision history. In our solar system, the orbits of all eight planets lie close to the plane that is perpendicular to the Sun's spin axis (i.e., the ecliptic plane), and all of the orbits are close to circular, except for Mercury, which has an eccentricity of about 0.2. An eccentricity of 0.0 means that an orbit is circular, and an eccentricity of 1.0 corresponds to an ellipse that is so elongated that it is a straight line.

Why Are Some Objects Known as "Candidate" Planets?

It is important, especially when reading or listening to news articles on exoplanets, that we are careful to distinguish between a *confirmed* planet and an object that is just a *candidate*. What is the difference, and why are these latter objects just exoplanet candidates? I will talk later about the details of how exoplanets are discovered and how physical measurements are made, but here it is sufficient to realize that there are many "false-alarm" possibilities. At a minimum we would want to observe more than one full orbit (the more the better) for an exoplanet candidate to be eligible for consideration for promotion to "confirmed" status. Obviously, if the period of revolution of the exoplanet around the host star is many months, the validation process can take years. In particular, if the exoplanet was discovered by the so-called "transit method" (in which the planet causes dimming of the host star when its path passes in front of the star), if you want to observe another transit, you have to wait for the duration of a whole revolution period. In general, the signals that are used to detect and measure exoplanets are weak and challenging to measure. They have to be pulled out from all kinds of other signals in the data ("noise" for this purpose), leading to possible uncertainties. The data have to be calibrated and checked for any other possible anomalies. On top of that, in the case of the transit method, it has to be shown or demonstrated that an occultation of the host star was indeed due to a planet and not another star. Star systems that consist of two stars in orbit around each other are known as binaries and can actually constitute most of the occultation events that are found. [9]

In addition to confirming transits, exoplanet candidates discovered by the transit method need to be followed up with additional observations with other instruments because the transit method itself does not yield all the physical parameters that are desired. Most notably, the exoplanet mass cannot be determined from transit observations alone. Even with additional follow-up observations, the exoplanet critical parameters such as mass are not measured directly, but by deduction using parameters pertaining to the host star. Even then, the desired quantities must be obtained by a combination of measurements and applying theoretical models (model-fitting), which may not necessarily yield unambiguous solutions (I will say more about this later).

In order to illustrate how careful you have to be with making a distinction between a *candidate* exoplanet and a *confirmed* exoplanet, I will illustrate with an example. Early in 2011, several results were announced from the NASA mission that was designed to search for exoplanets (*Kepler*). Over a thousand exoplanet candidates were found, and a six-planet system (with confirmed planets) was also found. A huge number of news articles appeared in all kinds of media as a result of the announcements. Now, a headline on one of the NASA *Kepler* website pages stated three different things in one sentence. It read, "NASA Announces 1, 235 Planet Candidates, Some in Habitable Zone, and a 6-Planet System." The headline is very unclear. It says that *Kepler* has so far discovered 1, 235 exoplanet candidates, some of which are in the so-called "habitable zone." The habitable zone is the region around a host star in which a planet can exist with a temperature and atmosphere that could harbor liquid water (and therefore presumably the planet could potentially support life). The third thing that the headline says is that *Kepler* has also found a six-planet system. However, the headline does *not* say that these six planets are in the habitable zone. The headline says that 1, 235 exoplanet candidates have been found by *Kepler*, and it is some of these *candidates* that are in the habitable zone. The six planets in the multiplanet system have nothing to do with the group of candidates because they are confirmed as being planets. The headline does not say that the latter six planets are in the habitable zone. I will say much more about the *Kepler* mission and about the caveats and problems associated with defining a habitable zone later in the book.

How Many Exoplanets and Alien Solar Systems Have Been Found?

The pace of discovery is so fast that the number of confirmed exoplanets sometimes changes daily. You can find out the latest numbers of discovered exoplanets and alien solar systems from the *Extrasolar Planets Encyclopedia* (`http://exoplanet.eu`). To illustrate the relative proportion of exoplanets versus alien solar systems, from a "snapshot" taken on the 8th of October 2011, there were 692 unique, *confirmed* planets discovered outside our solar system. They reside in 567 unique planetary systems, 81 of which harbor more than one planet. Since the numbers change so fast, it is impossible, in a book like this, to show you diagrams and figures that contain the results from all of the known exoplanets at any given time. At some point you have to "freeze" the data for analysis. Therefore, the figures and discussions in this book will be based on the sample of confirmed exoplanets that existed on the 8th of October 2011 (i.e., the sample with 692 unique exoplanets). The results in this book will be updated in the future to include exoplanets that were discovered after the cutoff date for this sample. In the remainder of this book, I will refer to the sample of 692 exoplanets as the "October 2011 sample."

Figure 1 shows a histogram depicting the number of exoplanets discovered per year (top panel), as well as the cumulative number of exoplanets discovered each year since 1989 (bottom panel). The numbers for 2011 pertain to the cutoff time for the October 2011 sample. I include in figure 1 the planets that were discovered that many scientists do not count as actual planets. These planets orbit objects that are not regular stars (but are "pulsars"), and the reasons for the controversy will be discussed later.

Most of the planets have been discovered by groundbased tele-scopes. The spacebased missions *Kepler* and *CoRoT* had, at the beginning of October 2011, discovered 24 confirmed exoplanets each.

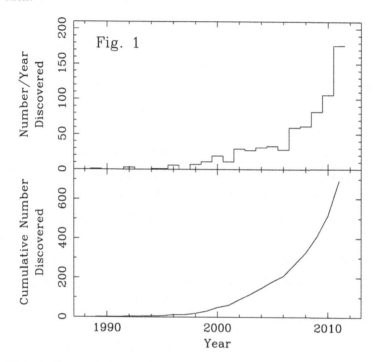

Figure 1: Top panel: A histogram showing the number of confirmed exoplanets discovered in each year since 1989. Bottom panel: The cumulative number of exoplanets discovered each year since 1989. Note that both graphs are complete only up to the beginning of October 2011. Data are from the *Extrasolar Planets Encyclopedia*.

Why Don't Exoplanets Have Names?

Exoplanets *do* have names, I hear you say. I'm sorry, but "2M J044144," or "1RXS1609" are not names. That would be like say-ing that the ISBN of a book is a name.

It is unfortunate that nearly all of the exoplanets discovered so far have extremely unimaginative names consisting of a bunch of letters and digits that only have a meaning to scientists (and even then, a very tiny group of scientists). One of the conventions used for exoplanet names is to use the host star identification (I hesitate to use the word "name") as the root and attach a lower case letter suffix to it for each exoplanet. A little bit better are the names that have been assigned to exoplanets discovered by the *Kepler* and *CoRoT* missions, which have the relevant mission name followed by an ordinal number that corresponds to the order of its discovery by the relevant mission. (More will be said about these missions later in the book.) This is hardly much better and no more imaginative than the "names" of exoplanets that were not discovered by these two missions. If more than one planet is discovered in a system by any of the two missions, guess what? The planet "names" are formed by appending a letter of the Roman alphabet to the root name (for example Kepler-11b). Why is it such an issue that the exoplanet "names" are boring and uninspiring? Should we care? I think it is extremely important because boring names turn people off. A layperson who may otherwise be keen and enthusiastic about learning about exoplanets and related science may be quickly put off after reading an article that is punctuated frequently by sterile and disruptive letter and number sequences. The reader's desire to learn may be cut short prematurely. This is a fact that is not appreciated by scientists who have let this happen, and it is not consistent with the goals of those same scientists (and the organizations that they belong to) of engaging, inspiring, and educating the general public.

I have heard the excuse that there are too many planets to name. This does not fly for two reasons. Firstly, there was a time when there were *not* too many planets but no planet naming was going on then either. Secondly, the *International Astronomical Union*

(IAU) keeps a list of over 16, 000 names of what are called "minor planets" in our solar system. That's shocking isn't it? Minor planets are small objects such as asteroids, which are knocking around in our solar system. Some of them are basically just big boulders. Examples of their names are things like "Agata," "Agnes," and "Zappa." If people can go to the effort of maintaining a list of thousands of objects that aren't even planets, surely it would take only a small fraction of their time to maintain a list of names of exoplanets that would not make a normal person run away. (There are a few exceptions: for example, the exoplanet HD 209458b has come to be known by the nickname *Osiris*.) What can you do about it? Express your opinion using every medium that is available to you, even if you think that nobody important is listening. Eventually, perhaps the message will be heard. You can easily find plenty of discussions and forums on the topic of exoplanet naming protests on the Internet. However, as you might expect, some of these forums attract some real "weirdos." (If you are one of them, it's okay, just remember to get out of the house once in a while!) There was one website I found that claimed to be dedicated to the exoplanet naming cause, but had huge threads on wedding dresses and very little on exoplanet naming.

What Do We Want to Know About Exoplanets and Why?

One of the ultimate questions we want to know the answer to is of course the "biggie." That is, do any of these exoplanets harbor alien life-forms, or are any of them capable of it? In order to even begin to answer that question, there is a long and arduous route consisting of many intermediate questions that must be addressed. I will come back to the question of life and the habitability of exoplanets in chapter 5, but for the moment I will pose some major intermediate questions that need answers.

For every exoplanet that is discovered, we would like to know its mass and size (radius). Together, these two quantities can be used to deduce an average density, which in turn can give us a crude indication of the chemical and physical composition of the planet. Is it made mostly of gas, or is it predominantly made of rock? Or is it mostly gaseous, harboring a rocky core? Or is it a so-called "ocean planet"? At present, the average density is the only indicator of the composition of most exoplanets, and despite the crudeness of the indicator, it is still useful. For example, if we find that the average density of a planet is similar to that of Earth we can deduce that the entire planet cannot be gaseous. Otherwise the average density would be much less than that of Earth (some of it could be more dense than Earth and some of it could be less dense than Earth, but it could not be *entirely* more dense or *entirely* less dense than Earth). I should add that determining the details of the composition of a planet is an area of largely uncharted territory, especially when we remember that we are not even certain of the nature of the very core of gas giants in our solar system.

We would like to know about the temperature (and its variability properties) of a planet. In particular, the surface temperature is critical in determining whether the planet can support life. For example, if water is a major requirement for life, then we demand a very strict range for the surface temperature for water to exist in liquid form. The surface temperature is determined by a number of factors, but it always comes down to the balance of heating versus cooling. Usually, the temperature of an exoplanet is estimated by drawing up an energy budget of all heating and cooling processes that we *think* are relevant, and applying theoretical models of those heating and cooling processes to come up with a temperature (this is typically an iterative, trial-and-error procedure).

In order to do that, the orbital characteristics of a planet must be measured. We would like to know the size and eccentricity of the elliptical orbit. The eccentricity of the orbit is a measure of how "squashed" the orbital ellipse is compared to a circle. The eccentricity obviously affects the distance of the planet from the host star (a significant source of heating). As a part of the process of mapping the orbit, we may learn about the inclination angle of the plane of the orbit relative to our sight line. This inclination angle is important for eliminating an ambiguity in the inferred mass of an exoplanet, but that angle is not often measurable. The time taken for one complete revolution of the orbit is also desired. To complicate matters, most of these "desired" quantities are not directly measurable. They don't just "pop out" from the data taken during the observations. They must be derived from the properties of the host star, which themselves are not all directly measurable. So the properties of the star itself (most importantly, mass and size) must be estimated as well. This is done by comparing theoretical models of stars with measurements of the stellar parameters that *are* accessible (including stellar temperature and luminous power) in order to come up with stellar mass and radius estimates.

You can see that there is a long chain of procedures that must be executed to be able to finally come up with the planetary properties that are desired. That chain carries along with it many assumptions, caveats, and uncertainties (in both theory and observation). What may not be so obvious when you read all the flowery news articles in the media is that even after going through the entire process, the mass of an exoplanet may *still* be unknown, aside from what is known as a "lower limit." A lower limit is the minimum mass that the planet could have, given all the observables and theoretical constraints. In fact, most known exoplanets only have a lower limit on the mass. (In the case of multiplanet systems, deriving mass constraints involves additional physics, is more complex, and can actually result in a mass *upper limit*, as is the case for an exoplanet called Kepler-11g.) You have to look at graphs and diagrams that include the exoplanet mass very carefully because usually even a lower limit on the mass is shown as a single point, hiding the fact that there is a large uncertainty with respect to the actual mass. Having said that, even though not knowing the absolute mass for an individual planet is a hindrance, when considering a sample of planets, it is possible to make certain statistical inferences because the unknown factor in the mass is related to the angle that the plane of the planet's orbit makes with the observer's sight line. The demographics and statistical information available from exoplanet observations will be discussed later. I will also say more about the actual mechanics of how different methods of exoplanet discovery, observation, and data collection lead to the deduction of various physical parameters later. We will see that no method is able to yield all of the parameters that are potentially accessible, and a hybrid method is often necessary.

One more thing that we would like to know is whether a particular exoplanet has an atmosphere and if so, what the composition of that atmosphere is. Answers to these questions have a direct bearing on what forms of life can be supported on the surface of the planet. Unfortunately, studying the atmospheres of exoplanets is still relatively difficult. The gaseous planets, even if they have a rocky core, are of course more amenable for atmospheric studies because they are mostly atmosphere in a sense, and they are generally larger. Indeed, the field of study of exoplanet atmospheres took off in 2002 with the gas giant known as HD 209458b, when sodium was discovered as a component of its extended atmosphere. This exoplanet (also known by the name of *Osiris*) has other claims to fame (aside from being unusual in having a "real nickname") so we shall encounter it on more than one occasion. Other elements have now been detected in the atmosphere of *Osiris*, including carbon and oxygen. [10] I will talk much more about exoplanet atmospheres later.

At a higher level of sophistication, it would be useful to be able to measure the rotational periods of exoplanets (i.e. the planet's "length of day," or the time taken for a planet to complete a spin on its axis). It would also be useful to determine whether the plane of the orbit is skewed with respect to the spin axis of the host star because a skewed orbit might give us clues about the conditions in the planet-formation stage of the early life of the system. Exoplanets do not emit their own (visible) light and the large distances of the planets from Earth result in signals that are very weak. Nevertheless, significant progress has been made, and more and more systems are turning out to have a so-called "spin-orbit" misalignment. Such a misalignment places severe constraints on the physical history of the system (formation, perturbation, and planet migration in particular). More will be said about this later. The angle between the plane containing a planetary orbit and the plane perpendicular to the spin axis of the host star is not to be confused with the inclination angle of the system (the latter measures the angle between the plane of a planetary orbit and the observer's sight line). The spin-orbit misalignment is most quantifiable for transiting planets, using something called the *Rossiter-McLaughlin effect*, which refers to particular Doppler distortions and modulations to signals received from a spinning object.

An important thing that cannot be done yet (again, due to the weakness of the signals) is the investigation of surface features of a terrestrial exoplanet, such as the presence of water, ice, craters, and volcanoes. Most exoplanets cannot be imaged yet, and the few that have been imaged have not yet been imaged with the necessary spatial resolution. At the beginning of October 2011 the *Extrasolar Planets Encyclopedia* listed 25 out of 692 exoplanets as having been detected by imaging, and I will say more about the direct imaging of exoplanets later. Intensive research and development to improve imaging sensitivity and capabilities continues.

How Far Away Are the Discovered Planets?

The nearest star to Earth is Proxima Centauri, at a distance of 4.22 light years. [11] None of the exoplanets discovered so far are located around this star or around the two associated stars, Alpha Centauri A and B. Figure 2 shows a histogram of the distances (from Earth) of all the exoplanets in the October 2011 sample that have measured distances. In that sample, the closest exoplanet lies at a distance of approximately 10.4 light years from Earth, and the farthest one lies at a distance of approximately $27,700$ light years from Earth. It can be seen from the distribution of Earth-exoplanet distance that the bulk (actually, nearly 78%) of the exoplanets lie at a distance of less than $1,000$ light years from Earth.

To put these numbers into context, you should remember that our galaxy, (the *Milky Way*, or *"The Galaxy"*) is very large compared to these distances. The Milky Way has a flattened, disk-like shape that looks like a pancake with a fat bulge in the middle. It also has spiral arms. The overall diameter of our galaxy is approximately $100,000$ light years. From a side view, the largest height (which is at the central bulge) is about $30,000$ light years. Our solar system is located at approximately a third of the distance from the center of the "pancake" to the edge, roughly at the midplane. In other words, it should be very clear that a significant fraction of the exoplanets discovered so far (about 78%) are located within the local neighborhood in our galaxy. The distance to

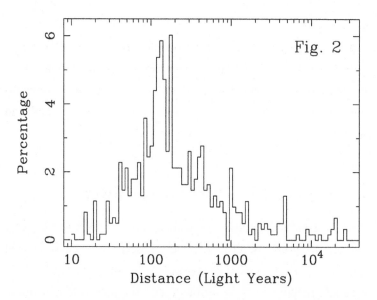

Figure 2: A histogram showing the distribution of the distance from Earth of the confirmed exoplanets in a subset of the October 2011 sample that have measured distances. Plotted is the percentage of the subsample of exoplanets in each distance interval. Data are from the *Extrasolar Planets Encyclopedia*.

the farthest exoplanet from Earth amongst this 78% (1,000 light years) is merely 1% of the largest dimension of our galaxy. There is an entire galaxy yet to explore, and an entire Universe of galaxies left to explore. [12]

An implication of the above distances to the exoplanets is that we are not talking about light travel-time delays of millions of years (as would be the case for exoplanets located in *other* galaxies). The maximum delay is about 27,700 years for exoplanets in the October 2011 subsample. In other words, the light from any of the exoplanets in the October 2011 subsample left the host star (or planet) less than 27,700 years ago. As implied above, in about 78% of the cases the light left its host star (or planet) much more recently than that (i.e., within the last 1,000 years). Therefore, most of these exoplanets are being observed in a contemporary epoch to ours.

A common question that many people have concerns the possibility (or otherwise) of humans ever traveling to an exoplanet, or sending a probe to one. I will talk about this later, with some additional technical details about issues relating to "time travel" and relativity.

How Many Alien Moons Have Been Discovered?

Moons, or natural satellites of planets, are common in our solar system. It should be clear from the preceding discussions that since exoplanets themselves are so hard to study because of the weakness of the signals from them, and the lack of direct imaging data, searching for exoplanet moons is even harder. However, one of the methods of studying exoplanets is in principle sensitive to the presence of a moon around an exoplanet (the so-called gravitational lensing method). Theoretical calculations have been done showing the expected signatures of exoplanet moons in the data. [13] However, this particular method of studying exoplanets is extremely labor intensive, and so far, at the time of writing, has only been used to study a dozen or so exoplanets. No moons have been found around these, or any other exoplanets.

You may wonder why it is important and/or interesting to look for and study exoplanet moons (often referred to as "exomoons"). One benefit of finding exomoons would be the possibility to obtain a better constraint on the mass of the exoplanet that the exomoon is orbiting (by means of measuring the orbital parameters of the exomoon). Also, moons are "worlds" in their own right and may be potential sites for supporting life, even though the parent planet may not be hospitable to life. Another reason why moons are interesting is that the origin of moons is still not completely understood.

In particular, the formation of moons has to fit in with any theory of planet formation, which itself is not understood. For example, in the case of Earth, did our moon form at the same time as Earth and become a separate entity, or was it captured? What determines if a planet ends up with a moon (or moons)? I will discuss the origin of our moon again later in chapter 5.

Are There Exoplanets That Are Not Associated with a Star?

So far we have talked about exoplanets that are members of alien solar systems. All of the confirmed exoplanets that have been discovered so far are associated with a host star. These planets are gravitationally bound to their host star in an orbit. However, it is important to realize that the methods of detecting exoplanets are in general predisposed to finding those that are gravitationally associated with stars. Remember that for most exoplanets, it is not possible to make direct measurements on the exoplanets themselves. Rather, the exoplanets are detected and studied by observing their host stars. Exoplanets do not emit visible light of their own so most of them cannot currently be directly observed. In the majority of cases, their presence is inferred from either their gravitational effect on the host star (producing "wobble" in the star's motion), or variable blocking of the light from the host star (causing "dips" in the observed light intensity from the star). In addition, an exoplanet may reflect light from the host star, and exploration of methods to study the (very weak) reflected light is something that is being actively pursued.

The question is then, is it in principle possible that exoplanets can exist without an associated host star, given that the current methods of studying exoplanets are biased against finding such exoplanets? Such "rogue" exoplanets would be wandering around in the darkness of space and would be very difficult to find. Nevertheless, the detection of such free-floating planets has been claimed. In particular, a study in 2011 found 10 free-floating Jupiter-mass objects that do not have a detectable host star within 10 AU. [14] However, the result is tentative. In the abstract, the authors of the paper very carefully used the term "Jupiter-mass objects" rather than "Jupiter-mass planets" (although popular news stories cried "planets"). It is possible that the host star is much further away than 10 AU. In fact the title of that paper uses the phrase, "Unbound or distant planetary mass population," but the razzamatazz of popular news reports that you will find all over the Internet marginalize the alternative, "distant," scenario, presumably because it is "boring" and not newsworthy. In addition, the abstract of the paper only says that the result "favors" the idea that the formation process of the wandering loners was different from that of stars and brown dwarfs. In other words, it cannot be ruled out that these objects are some kind of failed star. Still, the authors of the study extrapolate from the 10 objects found and speculate that they could be twice as common in our galaxy than the most common type of star. This is an extraordinary extrapolation considering that there are about 100 billion stars in our galaxy.

Whether the objects are planet-like or not, their formation process is uncertain. Could they really be planets that were ultimately ejected (scattered) into "unbound or very distant orbits?" Planets are conventionally thought to be formed from the material of a star, but our understanding of the planet-forming process itself is questionable. I shall return to the planet-formation problem on several occasions in this book.

How Can I Get the Latest Updates on Exoplanets and Alien Solar Systems?

Exoplanet research and the quest for life elsewhere in the Universe is an extremely rapidly advancing field. For researchers in the field and professional scientists in some other fields it is fairly easy to find the latest information, although it can be time-consuming sifting through the many research papers that are published every month. Although key information is distilled at various websites, the information is rather dispersed and heterogeneous. For the layperson the various Internet outlets can be confusing, and gleaning the correct message from popular news articles is not usually easy. There is a lot of "noise."

For the latest information on how many exoplanets have been discovered, with a breakdown of how many multiplanet systems have been found, the *Extrasolar Planets Encyclopedia* (http://exoplanet.eu) is a very comprehensive and reliable source. This website also carries databases that contain the most important actual measurements that have been made, with various useful breakdowns. People who are interested in using the actual data (for example, educators, researchers, and adventurous laypeople) can download the actual data at no cost. The database carries comprehensive citations for the original sources of measurements and results. However, the website is not exactly user-friendly, and the interested layperson (or even researcher in a different field) may not find it useful, especially as the meanings and caveats of some of the data columns may not be obvious. There is also a list of selected websites in appendix B.

Chapter 3

How It's Done

The Host Stars

As I mentioned in chapter 2, almost none of the actual physical quantities that we would like to measure for an exoplanet can be measured directly. Most of the parameters must be derived, and these derivations make use of the mass and size of the host star that the exoplanet is orbiting. The next problem is that we cannot simply look at a star through a telescope and measure its mass and size (radius). Most stars are too small in apparent size even when they are observed with a telescope. The mass of any astrophysical object has to be inferred indirectly using other quantities that *can* be measured, and applying the laws of physics. As for the size of the star, the laws of physics alone are not enough. We need to actually understand how stars work and then construct a model of a star using complex computer code to mimic what's going in the star. We plug in what we do know about a particular star and have the code predict what we don't know. There is a "catch 22" here of course, because we can ask how we know that the predictions are correct without testing them against a real star with real mea-surements of the predicted quantity. Well, it's a bit like squeezing a plastic bag that is full of air. If you squeeze one part of the bag, another part inflates more. If you squeeze that part, a different part of the bag gets puffed up. In other words, if the mass and radius of a star are too far off the mark in the theoretical model of that star, other properties of the star will be so much in disagreement with experimental measurements *of other quantities* that large regimes of mass and radius can be ruled out. Some of these other quantities for a star (for example, the surface temperature) can be measured with very high precision. In the end, the parameters that are inferred are a mass and a radius that each have a margin of error corresponding to how much each parameter can be "maneuvered" before other measurables are thrown off to contradict experimental values of things that *can* be measured.

So, in order to understand how the key quantities for exoplanets are measured or inferred, we need to understand how stars work. This is not a bad thing because stars are fascinating phenomena in their own right. Since the first observations that broke down starlight into its distribution over wavelength (i.e., since stellar spectroscopy became possible), over a 150 years of observational and theoretical work on stars has led to a robust understanding of their structure and evolution, at least during the phases of a star's life that come after the initial ignition of nuclear fuel burning. More details about the developments that were necessary will be discussed later. There are still some important unknowns and uncertainties in stellar structure theory, but I will not discuss them here. That's another book. However, stellar structure is understood well enough that computer models can be used to simulate stars with sufficient confidence to enable exoplanet parameters to be derived (albeit with sometimes significant margins of error).

In essence, we can think of a star as a fantastically immense chemical factory, making various elements using nuclear energy, starting from the simplest ingredient, hydrogen. A star starts its life and journey as a tenuous body of gas that is dominated by hydrogen, the most abundant element in the Universe (about 90%, counting by the number of atoms). If the body of gas is large enough, and if there is some disturbance that makes the gas depart from a smooth, uniform distribution, the gas collapses under its own gravity to form a compact object. This is an incredibly traumatic process because very high velocities are attained during the collapse and very high temperatures are created. At some point, by means of a process that is not yet properly understood, the collapsing object forms a pair of "spindles," (jets, in jargon), consisting of high-velocity material that is actually *outflowing*. These objects have actually been observed and are called protostars. The mechanism by which the extremely powerful jets form while the core object is collapsing is a fascinating mystery.

At some stage the temperature at the very core of the collapsing object becomes high enough (about 15 to 20 million degrees Celsius, or 27 to 36 million degrees Fahrenheit) to ignite nuclear burning which begins the process of fusing hydrogen to form helium, the second-most abundant element in the Universe. The actual temperature and other physical details depend on various initial conditions and the chemical composition of the constituent gas. Although the heavy elements have a much smaller abundance than hydrogen and helium, they nevertheless affect the physics of star formation. As well as having a high temperature, the core is clearly crushed by an immense pressure from the huge weight of the star bearing upon it. The jets vanish, again by a mechanism that is not understood, and the star assumes a stable spherical configuration in which it is steadily fusing hydrogen into helium at its core. The actual size and mass that is assumed by the star in the steady state is determined by the fact that the outward pressure from the nuclear burning must exactly compensate for the inward gravitational collapse that is trying to take place. The star then settles in for a long stretch of equilibrium in which it shines steadily, powered by nuclear burning at its core. The jargon for this phase is the "main sequence." During the main sequence phase, the mass, size (radius), surface temperature, and the luminous power output are not all independent because the steady state is determined by some very basic physics that does not allow enough degrees of freedom for all four quantities to be independent. Thus, not all four quantities need to be measured (and indeed cannot usually all be measured), and theory can be used to estimate one of the parameters from some of the others. Note that the surface temperature is much less than the core temperature (of the order of thousands of degrees, as opposed to tens of millions of degrees in the core). At the next level of detail, the theoretical relations between the four parameters are not completely straightforward because the final stable configuration does depend on the detailed chemical composition, so in practice full-up computer codes including all the physics have to be used to model the star and derive the required parameters.

From the point of view of stability, a star on the main sequence is very good news for an exoplanet, especially one that harbors life. Our Sun is on the main sequence and is about halfway through its stable hydrogen-burning phase, which is estimated to last a total of around 10 billion (10^{10}) years. The majority of exoplanets have been found, not surprisingly, around main-sequence stars, but later we will also come across exoplanets that have been found around more evolved stars.

So, given that we can confidently model a star on the main sequence and infer a mass and radius for the star, how can we tell that a particular star is on the main sequence? First, we should ask what we actually observe when we look at a star through a telescope. By breaking down the observed light into different wavelength (or color) regimes we can look for signatures of various elements present (and estimate their relative abundances), and we can measure the surface temperature from those data. In jargon, the term "spectrum" refers to the data that are broken down by color. The total power output can of course also be obtained from the data. Extensive statistical studies of stars over many decades have shown that if we place the stars on a diagram that plots power output versus temperature (which is related to an effective "color"), stars at different stages of evolution occupy different parts of the diagram, following well-established evolutionary tracks. A new star that is observed can by placed on this type of diagram, which then gives us important clues about what stage it is at in its evolution. (The diagram is known as a Hertzsprung-Russell diagram.)

After the hydrogen is exhausted, core helium production ceases and the central pressure drops because the nuclear fusing stops. The helium core then collapses under gravity and this dramatic event leads to a kind of back-reaction in which the outer parts of the star start expanding while the core collapses. The outer envelope is cool so it appears to be red, and this phase is known as the red giant phase. The size of the outer envelope can become so large that it can extend to smother one or more of its planets if it has any (this will be the fate of Earth when our Sun becomes a red giant in 5 billion years or so). Once the hydrogen is exhausted, the star has left the main sequence.

The collapsing helium core eventually reaches a high enough temperature to begin fusing helium into carbon. Thus, the life of a star continues. When the helium is exhausted, gravitational collapse ensues again until the temperature is high enough to fuse carbon. In this way oxygen, neon, sodium, magnesium, silicon, sulfur, and other progressively heavier elements are made until we

43

reach iron. To make further new elements beyond iron would require an input of energy rather than nuclear reactions producing energy. From here several evolutionary paths are possible, and I will pick up the story again later when discussing exoplanets around stars that are not on the main sequence. Note that only certain elements are produced by nuclear reactions in the cores of stars. Some elements that are not made in this way are made during so-called "supernova" explosions. There are a number of reasons why a star could explode in this way, but I will not discuss those here.

Before we go back to main sequence stars, it is important to realize that the gas that is expelled into the Universe during the course of stellar evolution can eventually end up somewhere else to take part in the formation of a completely new star from scratch. Such a star will already have some of the elements present that were made in a previous star and the new star will be a second-generation star. This is why the Sun, for example, already has elements present that are more complex than helium, even though the Sun is a main sequence star. Indeed, if planets form early in the life of a star the required elements already need to be present. By similar reasoning, there should also exist stars that are made purely of hydrogen and helium. However, it is a major puzzle in astrophysics that such first-generation stars have never been found.

We can see how measuring the relative abundances of elements in a star can tell us a lot about the stage of evolution that star is at. The very same atoms of elements that are made in the stars not only end up in planets, but of course can also end up in living creatures. For example, every atom in your body once resided in a very harsh environment in the core of a star. The atoms in the clothes you are wearing, and in the food that you eat, were all once participating in the unimaginably fierce nuclear reactions in the heart of a star.

How are Exoplanets Discovered?

Exoplanets were not discovered by accident. Humans have of course wondered for a long time about life elsewhere other than Earth, but we have to remember that it was not obvious to people that the objects in the sky that moved and were not stars were potentially habitable places. It was not even obvious prior to telescopic observations that they could be other worlds. We also should remember that the telescope came into use not much longer after the death of Copernicus (in 1543), who put forward the idea that the Earth is not at the center of the Universe and that the Sun does not revolve around the Earth. Galileo, who is associated with the earliest telescopic observations of heavenly bodies, championed the Copernican view but the matter was still highly controversial during his lifetime. Despite that, in 1610 Galileo discovered three moons of Jupiter using a telescope, although at first he did not realize the full impact of what he was seeing. He did manage to conclude that these tiny, newly discovered objects were orbiting Jupiter, which completely threw off the Aristotelian view that absolutely everything in the sky that moves goes around the Earth (a view which was still prevalent at that time). [1]

It was then just a matter of time for the necessary innovations and improvements in instrumentation to be realized, before people started deliberately looking for planets outside our solar system. Obviously, for this to happen, people had to realize that stars are analogous objects to our Sun. Although this idea had been thrown around in ancient times, humans were not in a position to begin to prove it until the nineteenth century, and before then there were plenty of competing ideas. Other books and sources will give names and dates to those in the ancient world who suggested that the Sun is a star, but I will give credit to the "unknown citizens," of which there must have been more than a few. I'm sure they did not all live in Greece either. The people recorded in history who are credited with the view that the Sun might be a star were no more qualified to make such statements than the "unknown citizen." The view was not supported by evidence or testable theories. In the nineteenth century it became possible to breakdown the light from stars and the Sun into many colors (in jargon, the result is a spectrum, and the activity is called spectroscopy). These studies, and the monumental task of classification of stars based on observational features, was a collective effort that spread across a period of over a century. Much of the classification was done before the physics of stars was understood so classification schemes underwent revision processes as the physics was gradually revealed. Notable amongst the first observers who observed the "decomposed light" (spectra) of stars and the Sun was Joseph Fraunhofer, in 1814. [2] Deducing the physics of stellar structure and evolution, and relating those findings to the observational properties and classification scheme of the stars is actually still an ongoing refinement process. The idea that nuclear fusion could account for the energy generation mechanism in stars was floated as early as 1920 by Arthur Eddington. However, there were many hurdles to clear before reaching the point of a satisfactory working theory of stellar structure and evolution that was driven by nuclear energy generation. By the late 1950s, the critical problems had been solved. [3]

Timing Method

Meanwhile, by the mid-nineteenth century, false alarms for the detection of exoplanets began, and continued right up to 1992, when the first bona fide exoplanets were actually discovered. The false alarms were due either to artifacts in the data and/or analysis, or misinterpretations of the data. The discovery of two exoplanets in 1992 was overshadowed by the fact that the host star was not anything like our Sun, [4] and in fact was a *pulsar*, which is a remnant of a star that exhausted all of its fuel for nuclear burning and collapsed until all of its atomic material was crushed and broken down to just neutrons. A pulsar is such a neutron star that is spinning, and its spinning magnetic field interacting with charged particles produces a powerful rotating radio signal that appears to be pulsed because the radio beam is highly directional. Since pulsars do not "shine," they do not produce light and heat as "regular" stars do, so it was assumed that the planets discovered around pulsars could not support life even in principle. However, this is not a scientific deduction that is based on evidence. It is an unproven proposition based on current knowledge and assumptions. We know very well that before the discovery of life in the deep oceans on Earth, it was generally believed that the existence of life at those depths in the ocean was impossible. Yet we now know that some deep-ocean creatures require no source of sunlight and thrive on the energy available from vents. The assumptions that supported the view that life could not be sustained without sunlight, and the deductions made from those assumptions (the prevailing wisdom, or knowledge) at the time, were clearly just plain wrong.

In the case of the exoplanets found around pulsars, the method of discovery involves measuring irregularities in the radio signal. Without a planet present, the periodic signature that is measured from the pulsar is extremely regular, stable, and clear. The time period between each pulse is easy to measure with a sufficiently high accuracy that irregularities due to the presence of a planet are predictable and measurable. Radio astronomy techniques were already in place and were being used in other ways before the detection of the exoplanets. The method is known (unimaginatively), as the timing method, and is applicable only to exoplanets around pulsars, not to exoplanets around "regular" stars.

The number of exoplanets discovered with the timing method is relatively small. At the beginning of October 2011, the *Extrasolar Planets Encyclopedia* reported a total of 14 exoplanets out of 692 that had the timing method applied to them. These 14 exoplanets reside in 9 star systems, 4 of which have more than one exoplanet.

Doppler-Shift Method

In the meantime it had easily been recognized that finding planets around regular stars was going to be extremely challenging because a regular star's light would so catastrophically swamp any reflected or intrinsic emission from the planet. One method of detection that seemed most promising was to look for the "wobble" in the position of the star itself, on the presumption that the orbit of a planet would cause the "wobble." This is expected because there are no "pillars" in space (or anywhere else for that matter) that hold things in a fixed position. Everything is relative. So a planet that is orbiting a star does not actually orbit a star that is stationary because there is no such thing as stationary without reference to another object. What happens is that the planet and star are in motion around their common "center of mass" point. The motion (with respect to the center of mass) of each of the objects in the orbit is commensurate with its mass. The higher the mass of the star, the smaller is the magnitude of the motion (or "wobble") of the star. On the other hand, a more massive planet would cause a larger magnitude of motion compared to a less massive planet. A planet that is closer to the star would also produce stronger variations than one that is situated further from the star. However, there is still a major problem: simple calculations show that the stars are so distant that physical positional wobble is generally too small to measure. On the other hand, the wobble is measurable by means of examining the variations in the wavelength (color) of light that is caused by the motion. This is the Doppler effect for light, analogous to the Doppler effect for sound, in which the pitch of a note changes with the speed of the emitter relative to the observer. For light, the relative motion of a star shows up as variations in the wavelength of specific "finger prints" that are unique to specific atoms. Those "finger prints" are like the bars on a bar code which can be compared with the bar code from the same atom observed in the laboratory (which should not have any bulk motion with respect to the observer aside from thermal motion). Comparing the two bar codes from the star and the lab may reveal a relative shift (offset) in the bars and that shift can be measured and converted into a relative velocity using standard physics. The velocity can be measured at various points in time and if the star is hosting an orbiting planet, the variations in velocity with time will show a specific form that depends on the various orbital parameters and properties of the planet and star. A model can be constructed with adjustable parameters. The parameters are adjusted to fit the data. This may not however, give completely unambiguous or unique solutions and the resulting uncertainties are limitations of the method, which nevertheless is very powerful.

The reason why the shifts in the wavelength are measurable whilst the positional variation of the star is not usually measurable is that the positional shifts in the lines in the bar code (in jargon, the spectrum) have nothing to do with the position of the star. The position of the lines depends on the relative velocity of the star, and in the instrument that is making the measurement, the positional shift in the lines is simply related to the optics and other parameters of the measuring device. By analogy, this would be like changing the ISBN of a book on a bookshelf. The position of the book on the bookshelf could remain unchanged but the bars on the bar code will be different because the ISBN has changed. The ISBN has nothing to do with the position on the bookshelf, just as the velocity shift of lines in the star's signature have nothing to do with the star's position in space.

Still, it should be realized that it was challenging to measure the small velocity shifts due to orbiting planets. A precision of the order of 0.01% or better is required. In absolute terms this corresponds to about 2 to 3 miles per hour, on top of actual speeds that are of the order of thousands of miles per hour (speeds in kilometers per hour are a factor 1.6 higher than those in miles per hour). For comparison, the speed of Earth around the Sun is a colossal 67, 000 miles per hour (approximately). The variation in this speed over a year in a complete orbit is the speed of the wobble that the Sun would be inferred to have if Earth were the only planet and the system was observed by distant aliens. Remember that only relative velocities have any meaning so 67, 000 miles per hour is the speed of Earth with respect to the Sun, and 67, 000 miles per hour is the speed of the Sun relative to the Earth. A 0.01% precision in this case would correspond to detecting variations of about 7 miles per hour, and these variations must be "pulled out" from the motion at 67, 000 miles per hour.

The above account is of course just a crude sketch of the actual process of measuring the velocity shifts of a star due to an accompanying planet. Many complicating factors have to be taken into account before arriving at the final answers, such as corrections for the motion of the Earth around the Sun. In other words the reference data in the lab are in orbit around the Sun so variations in the Earth's orbital velocity will affect the results.

During the 1980s and early 1990s a 12-year observing campaign was performed to look for exoplanets using the so-called velocimetry method described above but it ended in 1995 reporting negative results, with no discoveries of planets in a search involving more than 100 stars (not pulsars). At that time only two exoplanets had been discovered and those were orbiting a pulsar, not a regular star (as described earlier). [5] However, later in the same year, the disappointment was trumped by the announcement of the discovery of an exoplanet around a sunlike star. The planet has an orbital period of just 4.2 days and an estimated lower mass limit of about half the mass of Jupiter. The exoplanet is known as 51 Pegasi b, or 51 Peg for short. [6]

The velocimetry method is also known as the radial-velocity, or "RV" method because there is a major caveat with the velocity that is actually measured. That is, the Doppler wavelength shift that is recorded can only pick up that part of the velocity change that corresponds to motion that is directly head-on (either receding or approaching). If the motion is oblique with respect to the line joining the observer and the star (which it will be, in general), then the measured velocity will be less than the actual velocity of the star in the direction of its motion. In other words the method does not pick up the transverse portion of the velocity. In order to correct for that we would need to know the angle at which the orbital plane of the exoplanet is inclined with respect to our sight line. However, the method does not always yield that angle in the set of solutions and it must be measured by other means. This is often not possible so there are a significant number of exoplanets whose orbital inclinations are unknown. This is the fundamental reason why many exoplanets only have a lower limit mass measurement as opposed to an absolute mass measurement. In order to actually measure the mass we would need to know the inclination angle for

rbit. In some cases it is possible to use supplementary infor-
.... on gleaned by other observational methods but in some cases
it is not possible. The cases most amenable to measurements of
the inclination angle are those in which the exoplanet occults the
host star ("transiting planets").

You may be wondering how people decide in the first place
which stars to look at in order to search for exoplanets. There
is no guidance. You can decide to choose stars in a certain mass
range (for example, you could decide to narrow the search to look
at only stars similar to our Sun). However, nobody knows for sure
how to predict, of all the possibilities, what kind of star is going
to harbor an exoplanet. These searches are therefore rather hit-
and-miss affairs. You have to select the particular telescope that
you are going to use and you have to select your stars somehow.
A huge effort must be invested by many people. As the statistical properties of an evergrowing exoplanet population are studied,
it is obviously hoped that we may eventually understand which
properties of stars, if any, are good indicators of whether a star
is likely to be accompanied by planets. It is already known that
there is tentative evidence that exoplanets are preferentially found
to be associated with stars that have the highest concentration of
elements "heavier" (more complex) than helium.

Astrometry

Although it is extremely challenging to perform direct measurements of the actual physical movement of planets and/or stars in
the plane of the sky, it is not impossible. The technique is called
astrometry and entails measuring the time variability of the spatial
coordinates of the host star. Although the method involves measuring actual positions in the sky, the method is subtly different to
imaging because one is not dealing with the image of a planet. The

host star is a "dot" and the positions of the "dot" are tracked in the time domain. The astrometry method is not currently amenable to "blind" exoplanet searches because it carries a high overhead in terms of observatory time, and groundbased observations are subject to limitations imposed by Earth's atmosphere. However, such searches have been attempted, but have not turned up any positive results. [7] A future spacebased mission called *Gaia* is planned to launch in 2013 and will conduct systematic searches for exoplanets that make use of astrometry.

Whilst astrometry by itself cannot currently be used to discover exoplanets, it can be used to provide supplemental information for exoplanets that have already been discovered and observed using a different method. For example, the ambiguity in the planet mass from the velocimetry method can be mitigated by supplemental astrometric observations, although the measurement uncertainties can be larger than those from other methods. The *Hubble* space telescope and the *Hipparcos* satellite have been used to make astrometric measurements of previously known exoplanets. Spacebased observations such as these are not only free of the limitations of Earth's atmosphere, they provide a greater field of view, which means that calibration (reference) stars can be selected that are further away from the target and are therefore not swamped by the light from the target. Even with these current spacebased missions, astrometry is only possible with the star systems that are closest to Earth. The spatial resolution limit of the instrumentation sets a limit on the distance of a system that can be studied. [8]

At the beginning of October 2011, about 92% of confirmed exoplanets had velocimetry and/or astrometry measurements. Check the *Extrasolar Planets Encyclopedia* for the latest numbers.

Transit Method

Another possibility that was recognized as a means of potentially discovering exoplanets was the prospect of observing an eclipse of a star by a planet that the star is hosting. It would not be possible to literally see the planet crossing the disk of the star because the planet would be so small and dim compared to the star. However, such a transit event would block a tiny amount of light from the star during the period that the planet would pass in front of the star. This would manifest itself as a tiny dip in the measured power (luminosity) of the star. Experiments of this sort are again extremely challenging because the instrumentation detecting and monitoring the dip would have to be so sensitive that it can accurately measure changes in the star's luminosity that are less than 1%. That sort of sensitivity applies to large planets that are of the order of 10% of the size the star. For smaller planets, the sensitivity has to be correspondingly better. For example, an alien observing the Earth in transit across the face of the Sun would have to be able to measure a 0.01% change in the received power from the Sun.

We see that the transit method is already biased towards finding the largest exoplanets. There is another preselected condition. That is, in order to observe an eclipse at all, the plane of the exoplanet's orbit must be oriented so that it is observed edge-on, or nearly edge-on. If alien solar systems harbor planetary orbits that are randomly oriented with respect to an observer, then there is a quantifiable probability that a particular system chosen at random is favorable for observing a planetary transit. The size of the exoplanet compared to the host star also affects this probability. [9] On the one hand, the bias towards an edge-on orientation is a restriction, but on the other hand it actually tells you something that could not be measured using the Doppler-shift method. Remember that with the Doppler-shift method (velocimetry), the inclination angle of the orbit with respect to the observer remained unknown and had a direct impact on the mass measurement of an exoplanet, rendering it unknown, except for a lower limit. Having said that, the transit method by itself cannot, even in principle, yield the exoplanet mass. This is because it is a purely geometrical phenomenon. Only measurements relating to the dynamics of a system could possibly constrain an exoplanet's mass. However, combining the transit method with the Doppler-shift method for a given system provides powerful complementary information because the inclination angle from the transit method can be used to remove the ambiguity in the mass obtained from the Doppler-shift method.

The transit method alone does not yield the *absolute* size of the exoplanet either. It can only give the ratio of the size of the planet to the size of the host star. The size of the host star cannot be measured directly in most cases and must be estimated by applying models of stellar structure, as discussed earlier. Therefore, we can indirectly estimate the absolute size of the planet from the inferred size of the host star.

Figure 3 shows a simplified illustration of a dip in the stellar power output when an exoplanet transits a star. The slanted portions of the dip correspond to the ingress and the egress, periods when the exoplanet is in the process of covering the star and uncovering the star respectively. The time durations of the ingress and egress are directly dependent on the relative size of the exoplanet compared to the star. If the quality of the data is sufficiently high it may be possible to learn more about the physical structure of the exoplanet from the detailed shape of the dip when examined over time. For example, if the exoplanet has a thick atmosphere, as opposed to a hard edge, the ingress and egress signatures will have extra structure superimposed on them because of the different opacities of the atmosphere and the body of the exoplanet.

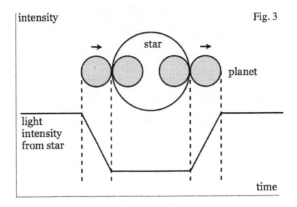

Figure 3: A simplified schematic illustration showing the principle behind detecting exoplanets using the "transit" method. As the exoplanet (small, dark circles) eclipses its host star, it moves from left to right in the diagram, reducing the observed intensity of the host star (as measured by a distant observer). The intensity of light from the system measured by a distant observer is then characterized by a dip (lower curve in the diagram). The size of the exoplanet relative to the host star and other factors (such as the opacity profile of the exoplanet) determine the precise shape of the dip.

If more than one eclipse is observed then the observations yield a direct measurement of the time duration (or period) of the orbit and this extra information can help to estimate the size of the orbit when combined with velocity information.

When the first exoplanet associated with a sunlike star was discovered in 1995 using the Doppler-shift method, the effort to find more exoplanets using that method was ramped up, and indeed more were found. At the same time, it was recognized that these same exoplanet data should be searched for transits across the host star. Having information from both methods for a subset of the

56

data would then yield a group of exoplanets with the best charac-
terizations of their physical parameters, especially the elusive but
all-important mass. The first transiting exoplanet was reported in
1999, and goes by the name of HD 209458b (an exoplanet that
has the nickname *Osiris*, and other claims to fame as well, so we
shall come back to it more than once). [10] It has a mass of about
70% of that of Jupiter and a size that is about 1.4 times larger than
Jupiter. The process of finding transiting exoplanets is slow and
painstaking, and in 2005 still only 10 transiting exoplanets were
known. That number had increased to more than 171 by early
October 2011 (check the *Extrasolar Planets Encyclopedia* for the
latest numbers).

To find completely new exoplanets using the transit method is
again an arduous task because it is not known where to look and
what stars are best to target. Portions of sky are observed and the
stars in the field are searched. Another major caveat of the transit
method is that intrinsic variations in the power output (luminosity)
of the star could be confused with a planetary transit. Remember
that the systems are generally too small to physically image the
transit. All that is observed is the power output of the star versus
time (so it is a "blind" observation in some sense). Other sources
of confusion are possible, such as another star crossing the sight
line.

Gravitational Microlensing Method

This formidable-sounding method makes use of the fact that light
bends when it passes near a massive object. Not only does light
bend, but the apparent brightness of the image received by a distant observer is affected disproportionately. The light bending and
amplification is an effect of relativity caused by the gravity of the
massive objects involved, in this case two stars. The details of

the effect are beyond the scope of this book, however. The two stars involved are not necessarily connected with each other, and one star could be much further away than the other. However, the light from one star passing closely to the other star on its way to telescopes on or near Earth carries signatures of the close passage that can in principle be calculated. The reason why this is relevant for exoplanet detection is that the signature contained in the observed power from the stars versus time can be profoundly affected if one of the stars is hosting an exoplanet. The exoplanet shows up as a very narrow but relatively strong spike in the brightness versus time domain. The method is also in principle sensitive to "free-floating" planets (i.e., those that are not associated with a host star).

As might be expected, so-called "microlensing events" are even rarer than transits, so the observations are even more difficult. There are many possible sources of confusion and error that must be taken into consideration, and the observing campaigns require huge amounts of observing time on multiple telescopes. It is a very labor-intensive effort. Since observing campaigns to study microlensing began in 1995, only 13 exoplanets were discovered by the beginning of October 2011 using this method, residing in 12 alien solar systems (check the *Extrasolar Planets Encyclopedia* for the latest numbers).

With such a low yield, and so many caveats, you may wonder whether it is worth all the effort. The unique thing about the microlensing method is that it is more sensitive to low-mass, Earth-like planets than any other method. It is sensitive to low-mass planets that are far enough away from the host star that water can exist in liquid and solid form (i.e. exoplanets that are beyond the so-called "snow line" or "ice line"). This gives considerable motivation for continuing to pursue the method because the other methods are not the best in this regime. Another advantage of the microlensing method is that it is very sensitive to multiplanet systems and it is sensitive to detecting moons around an exoplanet. The method is also sensitive to systems that are fainter than the ones that could potentially be detected by other methods because it does not rely directly on the light intensity of the host star. Moreover, it is not required to observe a complete orbit of the exoplanet with the microlensing method because the detection is virtually instantaneous. This is especially important for exoplanets that have orbital periods of many years. For example, for a planet like Jupiter that has a period of about 12 years, it would take that long to validate data using a method that required a complete orbit. On the other hand, microlensing events are one-time events. There is no possibility of repeating the experiment in order to validate any questionable findings in the data. This contrasts with methods such as the Doppler shift and transit methods for which data can be gathered from multiple orbits for confirmation and validation.

Direct Imaging

Compared to the methods discussed do far, direct imaging of exoplanets faces the most formidable hurdles, and extraordinary challenges lie ahead. Yet the method has the potential to return the highest bounty. Direct imaging is not to be confused with astrometry (discussed earlier), which maps the variable spatial position of the host star. With direct imaging, you actually see a dot that is the planet, and it is the light from that dot that is studied. In other words, the other methods make measurements on the host star but direct imaging makes measurements on the planet itself.

Even though what you see is a tiny dot, you can't help being overwhelmed by a powerful feeling when you look at that dot (which is very far removed from any of those "artists impressions" of alien planets that you see). The knowledge that you are looking at another dot that is *another world* is intoxicating. We have come so far as to be able to look at another world and even to be in a position to wonder whether there are alien "people" on it, living their alien lives, enjoying their alien Sun. Or is that other world home to gigantic and fabulous creatures as Earth once was to the dinosaurs?

With direct imaging of exoplanets it is possible to find out, in much more detail than other methods, about the chemical composition of the planet and its likely temperature. Signatures of chemistry from the planet's surface and/or atmosphere can be measured directly (in principle). This is why the potential return from exoplanet imaging is so high: discovering and accessing the signatures of alien life becomes a real possibility.

The observations are challenging because the projected distances of the planets from the host stars are so small and the images of the planets themselves are so small. Extremely high spatial (angular) resolution is required. The contrast in the light intensity of the star to that of the planet is extremely high and must be tackled. Scattering of light in the instrumentation must also be kept under control. In fact direct imaging is done not only with visible light but infrared light as well. The overall signal from the planet is a combination of reflected light from the host star and some emission from the planet itself (planets do emit their own infrared light). It is clear that the easiest systems to observe will be those that are closest to Earth and those that have planets that are farthest from the host star. Successful direct imaging has been performed for a limited number of cases, the first in 2008. [11] There are many current programs underway, and many under study to address the overwhelming technical challenges.

It is natural to wonder whether the enormous effort in direct imaging exoplanet research is worth it. I am impressed by some closing remarks by Traub and Oppenheimer in their very thorough and comprehensive (technical) review of the state-of-the-art of exoplanet imaging. [12] The remarks are very optimistic and remind the reader, in a way that is unusual in a scientific paper, of why the effort will continue. In talking about whether moving away from groundbased telescopes to spacebased telescopes is the logical choice for advancing exoplanet imaging techniques, Traub and Oppenheimer say, "Although we believe that this is true, one of the biggest mistakes a scientist can make is to assume that the primary mode of attacking a particular question is the only one." In other words, what they are saying is that there may be a clever solution or solutions that nobody has yet thought of. They then go on to make the rather moving remark that, "As such, one must remain optimistic, and in the end, the overwhelmingly compelling nature of the science of exoplanets, and how they directly relate to our own existence, means that the science will get done. Whatever mission, telescope, or technique is eventually used, perhaps even within the next 20 years, other planets similar to Earth with telltale signs of biological forcing of atmospheric chemistry will be discovered." Note that they do not say, "*might* be discovered," (as is more usual in a scientific paper), they say, "*will* be discovered."

That is what it's all about, folks. It doesn't matter how difficult the job is. We want to search for other life in the Universe so badly that it *will* get done. I don't know why we want it so badly. It is inexplicably fascinating.

At the time of writing, the fraction of exoplanets that have been amenable to study with direct imaging is only of the order of 4% of the total number of confirmed exoplanets. Check the *Extrasolar Planets Encyclopedia* for the latest numbers.

What Can We Actually Measure?

I will now pull together all of the different aspects of measuring observables of exoplanets so that you can see what we end up with and how that relates to what we want to know. This is important because when you read news articles in popular publications you will not always be told things that are not obvious and some of the things reported may be inaccurate or misleading. For example, in April 2011 there was much excitement in the media about the discovery of a "super-exotic super-Earth," making it the "densest known rocky planet." One of the press releases reported that the exoplanet in question was about 60% larger than Earth but eight times more massive, making it twice as dense as Earth. Such a high density would require somewhat extreme physical conditions. However, if you consult the scientific paper that reported on the same data (analyzed by the same people) for the same exoplanet, in a peer-reviewed journal, there is absolutely no mention of the unusual result reported in the popular media. In fact the abstract of the paper clearly reports that the size of the planet is *twice* as large as that of Earth and the density of the exoplanet is *similar* to that of Earth. What happened there? Well, clearly the radius of the exoplanet was subject to considerable uncertainty and in this case the initial estimate reported was wrong by a factor of the order of 25% and the press releases gave no hint of the possible uncertainties in the radius measurement. Those scientists had to know that the radius *could* be off by as much as it was because assessing such uncertainties is part of the scientific process. What is wrong is that the press releases did not acknowledge this possibility, giving numbers without any range in uncertainty. What makes this case even more of a disaster is that for a given mass, density is inversely proportional to the radius *cubed*, so getting the radius wrong by a factor of 25% gets the density wrong by a factor of two in this example. Unfortunately the scenario just described is not uncommon. There is way too much noise and hullabaloo in the way the results of scientific research are disseminated to the public. There is too much pressure to be sensational. Scientists tend to think that only superlatives will impress the layperson. Any talk or mention of uncertainties and, shock-horror, measurement errors, is deemed to be boring. A press release would never tell you that, due to measurement errors, it is possible that the sensational result reported in the press release may vanish. I don't think that the typical layperson is that shallow. So, whenever you read anything that sounds like a sensational or spectacular scientific result, you should always wonder what the numerous caveats are that are not mentioned. Obviously it is not always possible to know these things if you are not an expert, but if in doubt, you can always reserve judgment rather than believe something is true simply on the given information. I also recommend that you stay away from so-called Internet "content farms." These use automated programs to collect noise from multiple sources into one website. It is convenient but deafening. Sometimes common sense is very handy. I once saw some postings on an Internet forum that was ablaze over a press release that announced that the "gamma-ray burst mystery had been solved after several decades." If that were true, wouldn't there be major announcements that all the scientific missions and programs designed to study this problem will be decommissioned and many scientists will now have to find other funding and projects? There were no such accompanying announcements. Those programs are still going, and papers about the topic continue to be published. What was *actually* reported in that press release was that somebody had found one *possible* explanation of the mystery and the press release had obviously not made it clear that many people put forward *possible* explanations all the time, and have done so for decades and continue to do so. Proving that an explanation is *unique* and is the one that corresponds to reality is a different matter entirely. I suspect that whoever was responsible for that press release simply wanted some attention.

Now let's go back to the question posed at the beginning of this section, "What can we actually measure?" First of all, it should be clear by now that we cannot usually measure everything that we would like to measure. The second thing that should be clear is that no matter what method that is being implemented to investigate exoplanet parameters, we must first get a handle on the key host-star parameters because direct measurements on the exoplanet parameters are not possible. Exceptions to this are direct imaging observations of exoplanets. However, even in the case of direct imaging, we must know, or have an estimate of the host star's mass, in order to arrive at an estimate of the exoplanet mass. The third thing to realize is that not all of the parameters of the host star are directly measurable either. Notably, the three most important things we want to know about the host star are its mass, radius, and distance (to Earth). The latter is directly measurable, usually by the so-called method of parallax, which makes use of the apparent shift of an object against a "fixed" background, when viewed from widely separated observing positions. The mass and radius must be estimated by measuring the star's energy output, both total, and the characteristic energy output broken down by color (in jargon, the spectrum). The measurables are compared with theoretical computer models of stars (templates) in order to arrive at estimates of the mass and radius. So you can see there is a lot of room for both observational and theoretical uncertainties in the derived stellar parameters, which will obviously have a knock-on effect on derived exoplanet parameters.

Now, for the exoplanets themselves, the three most important things that we want to know are the mass, radius, and temperature. I haven't talked much about the temperature yet and will defer that discussion until later. Suffice to say, it involves even more of a convoluted path to arrive at an estimate of the surface temperature of a planet, and even then the estimated temperature may be off by hundreds of degrees. Yet the temperature is obviously one of the critical factors that will determine the possibility of the existence of liquid water, and (presumably) of life on the planet. The other parameters of interest for exoplanets are the orbital time period (the planet's "year"), the size of the orbit (as characterized by the maximum distance from the host star), the eccentricity of the orbit (how squashed the shape is compared to a circle), and the inclination angle of the plane of the orbit to our sight line. These parameters, if they are measurable, will usually "come out" as part of the overall process of determining the mass and radius of the exoplanet.

If the mass and radius of the exoplanet are determined, then we can get an estimate of the average density, and thereby begin to put some constraints on the chemical composition, as discussed earlier.

If the exoplanet is amenable to having its light analyzed (either direct infrared emission, or reflected starlight), then we can begin to put some serious constraints on the composition of the exoplanet and any associated atmosphere.

Knowing the distance of the exoplanet from its host star, and including some other assumptions about the reflectivity of the exoplanet, we could estimate a nominal temperature. However, without detailed knowledge of the atmospheric composition these temperature estimates could be way off (so the problem is a bit circular). Knowing the mass and radius, we can also estimate the surface gravity, and with the addition of a temperature estimate, we could put constraints on what type of atmosphere the exoplanet could hold on to. In other words, we may be able to rule out the existence of an atmosphere that consists of molecules that are so light (in weight) and so fast in speed (due to the temperature), that they would escape the gravity of the exoplanet (the molecular velocity would exceed the exoplanet's escape velocity).

So, starting with the most common method of measuring the parameters of exoplanets (velocimetry), and assuming we have measured the parameters of the host star, what do we get? What we get is a bunch of measurements of something that is not what we actually want to know, but a convolution of several things that we want to know, at a number of points in time. This quantity (in jargon, it is known as the "velocity semiamplitude"), when plotted against time is a curve that must then be fitted with a theoretical model in which the parameters that we do want to know are adjusted until the best fit is obtained. There is thus some "slack" to be expected, whereby a range in each parameter is allowed for a valid solution. Two additional factors that must go into the model are basically orbital reference parameters. One is a parameter that tells you where exactly in the orbit your data start (you won't know this beforehand), and if the orbit is not circular, you have to know how the orbit is oriented with respect to you. In other words, we are talking about the rotation angle (with respect to you) of a reference point on the orbit, such as the point of closest approach of the exoplanet to its star. This reference angle is manifestly a different quantity to the inclination angle of the plane of the orbit with respect to you. At the end, what you get is the eccentricity of the orbit, the time period for a complete revolution, the dimensions of the ellipse that corresponds to the orbit, and a lower limit on the exoplanet's mass. You do not get the actual mass because you do not get the inclination angle of the plane of the orbit, and instead you get the mass multiplied by something that depends on the inclination angle. On the assumption that, for a sample of exoplanets, the orientation of the plane of the orbit in the sky is random, you can calculate what this uncertain factor in the mass might be, in terms of a probability. It turns out that 87% of the time the uncertainty should be less than a factor of 2. In other words, you can say with 87% confidence that your actual exoplanet mass is less than double that which you deduce as a lower limit. Another thing that you don't get from all of the above is the exoplanet's size (radius).

If you find that, in addition to the above results, that your exoplanet also transits (eclipses) the host star then you can get additional information. The fact that the exoplanet crosses the face of the star at some point in its orbit already tells you that the plane of the orbit is observed edge-on or nearly edge-on, and you can get a good handle on the inclination angle of the orbit. You can then go back to the estimate of the lower limit on the exoplanet mass, plug in the inclination angle, and get the actual mass. From the transit observations you can get the ratio of the size of the planet to the size of the host star. Using your estimate of the host star's radius, you can then get the size (radius) of the exoplanet.

If the exoplanet is first discovered by the transit method then you do things the other way around. You measure what you can from the transit data and then do follow-up observations to get velocimetry data and "fill in the blanks," so to speak.

Oblique, Retrograde, and Eccentric Orbits

Ideally, we would also like to be able to measure an exoplanet's spin (or equivalently, the duration of its "day"). This is not possible at the moment. On the other hand it is possible, in some cases, to measure the spin rate of the host star. The rotation axis of the spin of the host star can be compared with the plane in which the exoplanet orbits. In transiting exoplanet systems, something called the *Rossiter-McLaughlin effect* can be used to infer this spin-orbit angle offset. An increasing number of exoplanets have been found whose orbital rotation around the host star is in the *opposite* direction to the rotation of the star (this is known as retrograde motion). Actually, retrograde motion is just an extreme case of misalignment between the orbital plane of a planet and the plane perpendicular to the spin axis of the host star (i.e., the misalignment is closer to 180 degrees than 0 degrees). As mentioned earlier, this has weighty implications for planet-formation and planet-migration scenarios, and I will say more about this later.

Chemical Composition

There is one very important measurable that I have not mentioned so far. That is what is known in the business as the "heavy metal abundance." No, this does not refer to the fact that your parents have a lot of vinyl records of dubious musical taste stashed away in the basement. Sometimes it is called the *metallicity* and it refers to the amounts of elements heavier than helium, relative to hydrogen. If you recall from the earlier discussion about stellar evolution, the relative amounts of carbon, nitrogen, oxygen, neon, and other elements all the way up to iron, are not only a smoking gun for the age of the star, but are also an indicator of whether the star is a first, second, or third generation star. Second and third generation stars are made from gas that was once part of a star in a previous generation that ended its life, exploded, and dispersed its material into space. The reason why the metallicity is important in the context of exoplanets is that it is becoming apparent that there may be a correlation between the likelihood of finding an exoplanet around a star with the metallicity (in the sense that higher metallicity indicates a greater likelihood), and I will say more about this later.

So, the question is, what has actually been done with the exoplanets discovered so far? Of the 692 exoplanets in the October 2011 sample, 686 have some sort of estimate for the mass, or lower limit on the mass (except for a few special cases which have *upper* limits on mass due to residence in multiplanet systems). Only about 28% of the exoplanets in the sample have a radius measurement. The corresponding (approximate) numbers for the size of the orbit, orbital period, and the inclination angle are 95%, 97%, and 27% respectively. About 72% have nonzero eccentricity estimates. About 89% of the exoplanets have a distance estimate. About 13% of the alien solar systems have no mass estimate for the host star, but about 49% of the exoplanets reside in systems that have an estimate of the heavy-element abundance (metallicity) in the host star.

To give you a rough idea of what information is known and what is missing for the less common methods of studying exoplanets, I will give some numbers that pertain to the October 2011 sample, and these serve as a rough guide on how the different methods compare. (Check the *Extrasolar Planets Encyclopedia* for the latest numbers.)

All of the transiting exoplanets in the October 2011 sample have a radius estimate, and all except 6 have a mass lower or upper limit. All except three of them have an eccentricity and an orbital inclination angle. All of the 13 exoplanets observed by microlensing in the October 2011 sample have measurements of mass and size of the orbit and a distance from Earth. Only half of these have an orbital period estimate, and only one has an eccentricity measurement. Two of the exoplanets, both in the same system, have an inclination angle measurement. None of the 13 exoplanets have a radius measurement and none have a metallicity measurement. These 13 exoplanets are amongst the most distant exoplanets known, lying between approximately 2, 000 and 20, 000 light years from Earth.

Of the 14 exoplanets observed by the timing method (in the October 2011 sample), all have measurements for a lower mass limit, the orbital period, and the size of the orbit. Only 10 of the 14 planets have an eccentricity measurement and only 2 have an orbital inclination angle. Out of the 9 systems containing the 14 exoplanets only 6 have a measured distance to Earth. Since the host star is a pulsar (a remnant of an expired regular star), metallicity in the usual sense does not mean anything.

Of the 25 exoplanets observed by direct imaging in the October 2011 sample, all of them have measurements of mass limits and the size of the orbit. Only 7 have radius estimates, and only 8 have orbital period estimates. Only 2 of the 22 systems of exoplanets have orbital inclination angle estimates, and only 5 of the 25 exoplanets have eccentricity estimates. The distance to Earth is estimated for 19 of the 22 systems containing the 25 exoplanets, and the metallicity is estimated for only 3 of the 22 exoplanet systems.

Now we come back to the sticky issue of the all-important exoplanet temperature. The estimates that appear in the literature are based on several assumptions. Firstly, using the temperature and power output of the host star, and the distance of the exoplanet to the host star, it is assumed that the luminous energy (which includes light), received by the exoplanet is partially reflected and partially absorbed, and that a steady state has been achieved in which the exoplanet is losing as much heat energy as it is gaining. Under these circumstances the temperature is stable by definition of equilibrium and can be calculated using basic principles of physics. However, this explicitly ignores any heat generation that is intrinsic to the exoplanet, and assumptions (actually guesses) have to be made about the fraction of the incident energy that is absorbed by the exoplanet (and therefore also the fraction that is reflected away). In jargon the parameter that expresses the fraction reflected is known as the albedo, but there are several ways to define the albedo because the geometry of the system and wavelength range of the light have to be specified. The procedure also explicitly assumes that there is no atmosphere. Usually there is no evidence that any of these assumptions are valid, and indeed, they may not be valid. For example, we know that Venus is way too hot for its distance from the Sun, due to thick layers of clouds in Venus' atmosphere. The predicted temperature according to the above method is about −40 degrees Celsius (about −40 degrees Fahrenheit), but the actual temperature is 462 degrees Celsius (864 degrees Fahrenheit). This is a whopping error of about 500 degrees Celsius (900 degrees Fahrenheit). You are also of course aware that the temperature on Earth can swing by 50 degrees Celsius (90 degrees Fahrenheit) or more for the same location on Earth, and it can vary by an even larger amount for different locations at a given time. The variations in temperature at a given location in time and place on Earth has very little to do with the variation in distance between the Earth and the Sun. The temperature has a complex dependence on many factors, including geometry, orientation of the spin axis of the Earth, and even the Sun's magnetic activity.

The lesson here is that you have to interpret exoplanet temperatures with extreme caution. If you see or read about an estimate of the temperature of the surface of an exoplanet, you have to first remember that the default estimate involves certain assumptions, and even if the reflectivity (albedo) estimate is reliable, you have to ask yourself how much is really known about the atmosphere and internal heat sources of the exoplanet in question. The *Extrasolar Planets Encyclopedia* and the *Exoplanets Data Explorer* do not give any estimated temperatures for the exoplanets. On the other hand, the website for the *Kepler* mission does give temperatures, and these are only *theoretical estimates* so they should be interpreted with caution. It is currently very difficult to study the atmospheres of exoplanets (more about exoplanet atmospheres later).

Groundbased versus Spacebased Observations

Most of the confirmed exoplanets discovered so far have been discovered and studied using groundbased telescopes that operate either in the visible light or infrared domain. Included here are the so-called "submillimeter" and "millimeter" domains, which have wavelengths longer than radiation in the infrared regime. The *Extrasolar Planets Encyclopedia* gives a long list of programs currently underway. Spacebased missions such as the Hubble Space Telescope (*HST*) and the infrared mission called *Spitzer* have been used in confirmation observations of a few exoplanets (details can be found in the *Extrasolar Planets Encyclopedia*). Although groundbased programs are still viable and productive, they are not ideally optimized to find the small, rocky Earthlike planets that are highly sought after. The two dedicated spacebased missions, *Kepler* and *CoRoT*, were specifically designed to search for Earthlike planets. They overcome the limitations (light "bending," extinction, and variability) due to Earth's atmosphere, the limited field of view, and the limited sensitivity of groundbased instrumentation. Both spacebased missions search for new exoplanets using the transit method, and are described below. In addition, the

Kepler mission provides *continuous, uninterrupted* monitoring of stars, increasing the chances of detecting an exoplanet transit in a blind search.

The CoRoT Mission

The rather dull acronym *CoRoT* stands for "Convection, Rotation, and Transits." The first two words refer to the fact that the mission is designed to do more than study exoplanets. It is designed to study certain aspects of stars as well, regardless of whether those stars are hosting exoplanets. Although many aspects of stars are now well understood, there are other aspects that are not. For example, the Sun, and other stars pulsate in many complex modes, and the physical origins of these pulsations are not well understood. This field of study is known as stellar seismology and has things in common with seismology on Earth, inasmuch as stellar rotation, pulsations, magnetic activity, and "starquakes" can give clues about the internal structure of a star. The term "convection" in the name refers to the fact that hot material in a star is cycled as it rises to the surface and falls back, in analogy to boiling water in a pan.

CoRoT is a French-led European mission that was launched in December 2006 and is expected to fly until 2013. The satellite is in orbit at an altitude of about 900 kilometers (560 miles) and has pointed at several locations in the sky in "blind" searches for exoplanets. As mentioned earlier, there will be many more occasions that are flagged as possible exoplanet detections because of occultations by another star (and not by a planet), or because of the intrinsic variability of a star. Recall that the change in brightness of a star due to an Earth-sized object eclipsing the star is tiny, of the order of 0.01%. By the beginning of October 2011, *CoRoT* had found 24 confirmed exoplanets (check the latest numbers at `http://smsc.cnes.fr/COROT/`). The masses of these exoplanets lie in the range of approximately 5 Earth masses to 22 Jupiter masses (about 7,000 Earth masses).

The Kepler Mission

NASA's *Kepler* mission was launched on the 6th of March 2009 and was designed to specifically look for Earthlike exoplanets around solar-type stars using the transit method. Such a dedicated spacebased mission overcomes the limitations of ground-based observations, as discussed above, but *Kepler* has one additional unique feature compared to the *CoRoT* mission. That is *continuous, uninterrupted* monitoring of stars in a blind search, increasing the chances of finding exoplanets transiting their host stars. In order to achieve the continuous monitoring, *Kepler* orbits the Sun, not Earth, and the orbit and mode of operation specifically avoids the Sun and the Moon getting in the way. *Kepler* trails just behind Earth in its orbit around the Sun.

The observing strategy of *Kepler* is that it simply stares at the same region of sky throughout its nominal 3.5-year mission lifetime (which could be extended up to 6 years). This region of sky is about 105 square degrees in size, constituting about 0.255% of the total sky area. The region is located near the constellations of Cygnus and Lyra, and was chosen to contain a high density of stars, again increasing the chances of finding exoplanets. The *Kepler* website (see appendix B) shows a visual map of the region of sky observed by *Kepler*. The advantage of observing the same region of sky is that multiple transits can be observed for a given exoplanet candidate. This is important, because, as discussed earlier, false alarms due to other stars eclipsing the target, or intrinsic stellar variability, need to be eliminated in order to promote an exoplanet candidate to a confirmed exoplanet status. For an Earthlike planet in the habitable zone of a solar-type star, the orbital period is obviously going to be about a year, so during the mission lifetime one could hope to observe three or more transits of a given exoplanet candidate. Three transits is the minimum number that is accepted by the *Kepler* team for rejection of a false positive.

The workhorse instrument aboard *Kepler* is what is known as a photometer, which is basically a light bucket. The larger the area of the light-collecting device, the higher the sensitivity and the better the signal strength compared to statistical noise. The photometer has an aperture of about a meter, and works with a telescope that has a mirror of diameter 1.4 meters. The sensitivity required to detect the transit of an Earthlike planet corresponds to a change in the measured intensity of the host star of 0.01%, which is incredibly small.

Over 150, 000 stars in the *Kepler* field are monitored, and they lie at distances of a few hundred to a few thousand light years from Earth. If a star hosts an Earthlike planet, the probability of observing a transit (given random orientations of the orbital plane) is only 0.5%. For Jupiter-sized planets the probability is correspondingly larger. The number of candidate exoplanets found by the beginning of October 2011 in the *Kepler* field was 1, 235. In addition to these candidates, there were 24 confirmed exoplanets up to that time, and none of them were deemed to be Earthlike. I will say more about the *Kepler* candidates and the confirmed exoplanets later. Check the *Kepler* mission website for the latest numbers and information.

When Was the First Exoplanet Discovered?

This question has not been answered earlier because the answer is actually not very simple. In 1989 there was a report of an "unseen companion" to a solar-type star, based on velocity fluctuations. [13] The lower limit on the mass was high, about 11 Jupiter masses, so it was concluded that the "companion" was probably a brown dwarf, although the abstract of the original paper stated that it "may even be a giant planet." As mentioned earlier, a brown dwarf is an object that could have been a star but is not massive enough to initiate nuclear reactions fusing hydrogen to helium. The lower mass limit is thought to be around 12 to 13 Jupiter masses, the mass threshold for deuterium ignition, and objects that have a mass greater than about 80 Jupiter masses are stars (see appendix A). The system has been studied repeatedly since 1989. As discussed earlier, there is a bit of a grey area between a very large planet and a brown dwarf. The *Extrasolar Planets Encyclopedia* now lists the object (HD 114762b) as an exoplanet.

The next exoplanet discoveries came in 1992, with three planets found around pulsars. Pulsars are remnants of stars that have exhausted their fuel and taken a specific evolutionary path. They do not therefore radiate like regular stars, which sets apart the exoplanets associated with pulsars. They are exoplanets nevertheless. The next exoplanet discoveries came in 1995, using the velocimetry method, after which the field really took off. The first transiting exoplanets were discovered in 1999.

Chapter 4

The Discoveries

Exoplanet Types

In the media and in the scientific literature you now see an ever-growing list of names referring to different types of exoplanets. You have your "hot Jupiters," "cold Neptunes," "mini-Neptunes," "super-Earths," "ice giants," "gas giants," and more. It is already terribly confusing. The problem is that the new discoveries demand an understanding of how planet formation gives rise to these different types of planets in order to come up with a classification scheme that reflects something fundamental about the structure of the planets. That understanding remains elusive so the classification of exoplanets is currently somewhat *ad hoc*. There are no quantitative definitions. This is common in any new field that uncovers objects that have not previously been studied. The same thing happened with stars. The classification scheme of stars underwent significant revisions before the current scheme was arrived at, and that scheme is founded on a physical understanding of how stars work. [1]

Instead of listing all the different permutations of exoplanet types, I will give you the information you need to figure out what a given exoplanet type is likely to mean, because there usually is some physical motivation behind the name. This will help to navigate your way around articles written about exoplanets.

A terrestrial planet generally refers to one that is rocky, like Earth, Mars, Venus, and Mercury. It is generally thought that the size of a terrestrial planet cannot be much larger than a couple of times the size of the Earth, and certainly not as large or as massive as Uranus, Neptune, Saturn, and Jupiter. This is not a fact. It is speculation based upon theoretical simulations that involve many assumptions and "hand-tweaking" (the problems with planet-formation theories will be discussed again later). As I will show below, of the exoplanets for which a mass *and* a radius are available, there are clearly a small group of planets that have an average density larger than Earth's and are hundreds of times more massive than Earth (with radii that are more than 10 times that of Earth). The terrestrial planets in our solar system all have a typical average density of around 5 grams per cubic centimeter (the latter volume is approximately equal to that of a typical sugar cube). This is to be compared with the density of water, which is around 1 gram per cubic centimeter, and with the giant planets in our solar system, which have average densities in the range of about 0.7 to 1.6 grams per cubic centimeter.

Whenever the name of a planet in the solar system appears in the name of a type of exoplanet, as you might expect, the implication is that there is some similar characteristic that is shared by the exoplanet and the planet in the solar system. Usually this refers to a similar composition, but it could refer to mass and/or size. It is very vague and fuzzy. Work associated with the *Kepler* mission adopts some very specific terminology for benchmarking the sizes of exoplanets and exoplanet candidates. Jupiter-size is defined as 6 to 15 Earth radii (Jupiter's actual radius is about 11 times that of Earth, and its mass is about 318 times that of Earth). Neptune-size is defined as 2 to 6 Earth radii (Neptune's actual radius is about 4 times that of Earth, and its mass is about 17 times that of Earth). Super-Earths are defined as having a radius between a factor of 1.25 and 2 times that of Earth. Anything smaller is termed Earth-size. However, these divisions are rather arbitrary and are by no means universally adopted. They mainly serve the purpose of placing exoplanet parameters into some sort of familiar context.

Prefixes like "giant," "super," or "mini" could refer to the mass and/or size of an exoplanet compared to a named solar system planet. For a given density there is obviously a correlation between mass and size. As we shall see below, the masses of the discovered exoplanets cover a much larger dynamic range than the radii, although we have to be wary of selection effects.

The prefix "hot" usually refers to a planet's relative distance from its host star. If a planet has no internal heat source of its own and no atmosphere, then its temperature is affected by its distance from the host star (and some assumptions about how incident radiant energy is reprocessed by the exoplanet). Thus, if two planets are similar, the planet-star distances are used as proxies for the surface temperatures of the planets. We can mathematically balance the heating from the star with the heat loss from the planet to come up with a so-called "effective equilibrium temperature." The calculation must make use of the host star's properties, as well as some properties of the planet that may have to be estimated (in particular, the reflectivity, or *albedo*). Although this calculated temperature may have nothing to do with the actual temperature of the planet, it serves as a quantitative "benchmark." Obviously, the further a planet is from its host star, the lower is the effective temperature. Thus, the prefix "hot" in the name of an exoplanet type refers to the fact that the planet is in some sense closer to the host star than was expected before that type of planet was discovered. For example, a "hot Jupiter" is a planet with a mass ("fuzzily speaking") of the order of, or larger than, that of Jupiter (Jupiter's mass is about 318 times that of Earth), but is closer to its host star than Earth is to our Sun. The effective temperature of a hot Jupiter is higher than the effective temperature of our Jupiter.

Extending the idea of effective temperature versus distance from the host star further, we can map the boiling points and freezing points of various elements and substances in terms of the distance from a host star because the effective temperature maps with distance. Again, this is only a baseline guide because the effects of an atmosphere and internal heat sources can profoundly affect the actual temperature of a planet. Nevertheless, for a given host star it is possible to calculate an estimate of the distance from the host star beyond which liquid water cannot exist due to the effective temperature being too low. The critical distance from the host star is called the snow line, or the ice line. You will see some exoplanet types with the prefix "ice" in the name, and this refers to the fact that the planet is further from its host star than the relevant ice line, so it is probably composed of ices. The ice may not be just water ice, but there may be other frozen substances such as methane and/or ammonia. It is not known whether the "ice giants," as they are called, actually have a rocky core as well. Recall that it is not known whether the gas-giant planets in our *own* solar system have a rocky core either. The "ice giants" are likely to have a hydrogen/helium atmosphere like Jupiter and Saturn.

In contrast to the prefix "ice," the prefix "gas" in the name of an exoplanet type refers, as you might expect, to the fact that the temperature is high enough for the composition of the outer part of the planet to be gaseous. You may be wondering how close a planet can get to its host star before it loses all of its water (if it had any) and/or other volatiles. It has been suggested that a so-called "ocean planet" (or "sauna planet") that is all ocean (with a depth of 100 kilometers (60 miles) or so) and steam, might take as long as 5 billion years to lose all of its water even if the planet-star distance were only 4% of that between the Earth and the Sun. [2]

Finally, I will clear up the term "Earthlike" planet. Even this term is used a bit too loosely in the literature and in the popular media. The word "like" itself begs for vagueness. Strictly speaking, "Earthlike" of course means that the mass, radius, and composition of the exoplanet in question are similar to those of Earth, and that the exoplanet is in the habitable zone of its host star. The vagueness enters when one questions what exactly is meant by "similar." Nobody expects to find a planet that is *exactly* identical to Earth and it would not be useful to define "Earthlike" as such. Nevertheless, the JPL (Jet Propulsion Laboratory) *Planet Quest* website home page has a counter that indicates the running total number of exoplanets that have been confirmed, but it also has a counter which says "Earthlike planets." The latter counter reads zero at the time of writing. There is no definition of precisely what "Earthlike" means, but there is a counter for it so some researchers must have made up some criteria that they will use to decide when the counter should be incremented. Presumably, once a few "Earthlike" planets have been found, people will start to think about (and argue about) how similar in mass, radius, composition, and temperature a planet has to be, to be called "Earthlike." Right now, nobody really has to deal with this awkward question.

Demographics

When considering the demographics and statistical properties of the discovered exoplanets and associated solar systems, you should bear in mind that selection effects, or biases, could be at play, and some of these are complex. In particular, the sensitivity of different methods of detection and observation may be different for different parameters. The principal biases for the two most common methods of observation (velocimetry and the method of transits) are well understood. For both methods, the sensitivity drops off with increasing orbital period of a planet. For velocimetry this is because longer observation times are required to accumulate a complete orbit, and these are more difficult to achieve for orbital periods of months to years. For the transit method, there is a bias against long orbital periods because the chance of observing a transit is obviously higher for smaller orbital periods.

For the velocimetry method, there is a bias against very eccentric orbits, and orbits that are observed close to face-on (perpendicular to the plane of the orbit). This is because in these regimes, there may be long periods of time when the velocity modulation versus time shows little variation because the most highly variable part of the curve is concentrated into a shorter period of time. The method relies on the ability to measure small variations in the velocity so long periods of little change will introduce bias because there will be fewer measurements of highly eccentric and face-on systems.

The amplitude of modulation of the velocity versus time curve will be larger for larger planet masses and it will be larger for planets that are closer to the host star, because both the mass and the planet-star distance obviously affect the motion of the star. Thus, the velocimetry method is biased towards massive, "close-in" planets.

The transit method favors finding planets that are large compared to the size of the host star because it relies on the ability to measure tiny dips in the brightness of the host star. The larger the planet, the larger will be the dip in intensity of the host star during occultation. The transit method is therefore biased towards planets that are large in size compared to the host star. The transit method is "blind" to the mass of a planet because the mass does not directly affect the probability of a transit and the mass cannot be extracted from the eclipse parameters alone. Therefore the transit method is *not directly* biased towards the masses of exoplanets. Of course, we know that many massive exoplanets are also very large and orbit close to the host-star, and these will be favored over small, less massive planets with larger orbits.

In the following sections I will talk about results from the October 2011 sample of exoplanets that was described earlier.

The Alien Solar Systems

About 89% of the exoplanets in the October 2011 sample had measurements of the distances to their host stars, and those distances lie in the approximate range of 10.4 to 27,700 light years. The distribution of those distances was shown in figure 2. It can be seen that the peak of the distribution lies in the approximate range of 100 to 200 light years.

The masses of the host stars in the alien solar systems lie in the range 0.02 to 37.5 times the mass of our Sun. However, the bulk of the stars have a mass that lies in the range 0.6 to 1.6 times the mass of our own Sun. This can be seen in figure 4, which shows the distribution of stellar masses. The strong peak around 1 solar mass, and the narrow width of the distribution is not surprising because stars on the main sequence, with a similar mass to our Sun, are deliberately targeted. There are other effects that complicate the selection biases for the host stars and each observation campaign has its unique set of biases. In particular, this means that we cannot simply take the number of planets discovered and divide by the total number of stars that were searched in order to estimate the frequency of occurrence of exoplanets. I will discuss this again later.

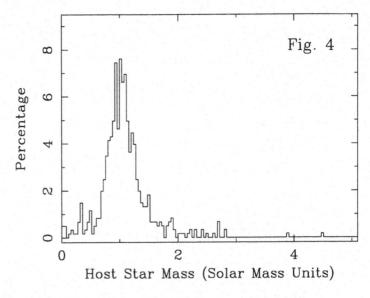

Figure 4: A histogram showing the distribution of the host-star masses associated with the confirmed exoplanets in a subset of the October 2011 sample that have measured stellar masses. Plotted is the percentage of the subsample of exoplanets in each mass interval. The stellar masses are shown in units of the mass of our Sun (i.e., solar mass units). Data are from the *Extrasolar Planets Encyclopedia*.

Masses of the Exoplanets

Over 99% of the exoplanets in the October 2011 sample have mass estimates. Remember that these mass estimates are lower limits (except for four cases in multiplanet systems that are upper limits), because of the uncertainty in the orbital inclination angle, so the masses could be a factor of up to about 2 larger, possibly more. In figure 5 a histogram of the distribution of these mass lower limits for the sample is shown, with the exoplanet masses expressed in units of Earth masses. The mass lower limits have not been corrected for the cases in which the orbital orientation is known. The inclination angle is known mainly for transiting planets (and for those the correction is small) so including such a correction would make the histogram confusing because it would be displaying a heterogeneous quantity.

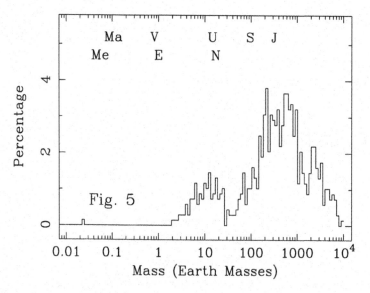

Figure 5: A histogram showing the distribution of the mass lower limits of the confirmed exoplanets in a subset of the October 2011 sample that have measured mass lower limits. Plotted is the percentage of the subsample of exoplanets in each mass interval. Masses are shown as ratios of the exoplanet masses to the mass of the Earth (i.e., masses are shown in units of the mass of the Earth). Data are from the *Extrasolar Planets Encyclopedia*. For comparison, the masses of the planets in our solar system are marked as *Me, V, E, Ma, J, S, U, N*, corresponding to Mercury, Venus, Earth, Mars, Jupiter, Saturn, Uranus, and Neptune, respectively (numerical values are from solarsystem.nasa.gov).

We see in figure 5 that there is a broad distribution of masses, with the bulk of the masses lying between about 100 to 10,000 Earth masses. For comparison, the masses of Mercury, Venus, Mars, Jupiter, Saturn, Uranus, and Neptune are marked at the appropriate positions on the plot with the letters, Me, V, Ma, J, S, U, and N respectively (obviously, the Earth has a mass of 1 in these units). The most massive planet in our solar system is Jupiter, which has a mass of approximately 318 times that of Earth. We see that the exoplanet mass distribution peaks between about 200 to 1,000 Earth masses, or about two-thirds to three Jupiter masses.

The exoplanet mass distribution also shows another small group of planets centered around 10 Earth masses, between 2 to 30 Earth masses or so. There is an outlier that is way off that has only 2.2% of the mass of Earth and this is one of the exoplanets found around pulsars (compact, spinning remnants of dead stars).

Exoplanet Sizes

Next, we can examine the size distribution of the exoplanets. Of the 692 exoplanets in the sample under consideration, only 192 actually have size (radius) measurements. The distribution of the radius (in units of Earth radius) for this subset of exoplanets is shown in figure 6. The percentages shown for each radial interval refer to the percentages out of the 192 planets with radius estimates and not to the full sample. We see that the bulk of the 192 exoplanets have a radius in the range of 10 to 17 Earth radii. The radii of Mercury, Venus, Mars, Jupiter, Saturn, Uranus, and Neptune are marked at the appropriate positions on the plot with the letters, Me, V, Ma, J, S, U, and N respectively (obviously, the Earth has a radius of 1 in these units). The largest planet in our solar system is Jupiter, which has a radius of about 11 times that of Earth. It turns out that the mass distribution of the subset of 192 exoplanets is similar to the mass distribution of the full sample, still peaking between 200 to 1,000 Earth masses. So the subsample with radii estimates is still dominated by giants. This is not surprising, because most of the exoplanets with size estimates are transiting planets and the transit method is "blind" to planet masses.

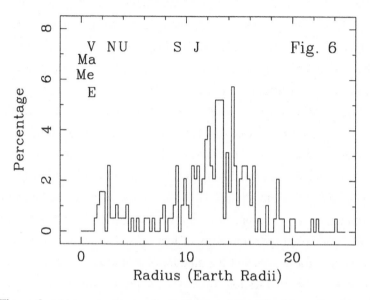

Figure 6: A histogram showing the distribution of the radii of the confirmed exoplanets in a subset of the October 2011 sample that have measured radii. Plotted is the percentage of the subsample of exoplanets in each radius interval. Radii are shown as ratios of the exoplanet radii to the radius of the Earth (i.e., radii are shown in units of the radius of the Earth). Data are from the *Extrasolar Planets Encyclopedia*. For comparison, the radii of the planets in our solar system are marked as *Me, V, E, Ma, J, S, U, N*, corresponding to Mercury, Venus, Earth, Mars, Jupiter, Saturn, Uranus, and Neptune, respectively (numerical values are from solarsystem.nasa.gov).

However, the question is, are the large planets also the most massive, or are there massive planets that are not giants? Alternatively, are there giants that are not massive? To answer this question we need to look at the mass and size distribution simultaneously. Remember, we can only do this with the subset of 192 exoplanets that have size estimates. In figure 7 the exoplanet radius (in Earth radius units), is plotted against the exoplanet mass lower limit (in Earth mass units). Also shown on the plot are the positions of our solar system planets for direct comparisons. We can see that, roughly speaking, for exoplanet masses and sizes less than Saturn, the larger the size of the planet is, the larger the mass is. This roughly translates into the fact that for low masses (less than about 100 Earth masses), adding mass to a planet makes a planet larger, and gradually less dense. Our familiar solar system planets lie on this basic trend. The average density of Earth is about 5.5 times that of water, and the density of our solar system planets decreases as we go out from the Sun, reaching a minimum value of about 70% of the density of water for Saturn. (The density of planets beyond Saturn increases again.) As we go from Mercury out to Jupiter and on to Saturn, the planets change from being rocky to gaseous. The giant gas planets beyond the asteroid belt, composed predominantly of hydrogen and helium, may also contain a rocky core but the amount of matter in the rocky cores remains an open question. Likewise, the proportion of the mass in a rocky core is also unclear for the gas-giant exoplanets.

The situation with respect to the density and size of exoplanets is completely different above about 100 Earth masses. As more mass is added, the exoplanet size does not increase any more and appears to remain in the approximate range of 10 to 20 Earth radii. This means that as we "turn up" the mass, the planet becomes denser again. The reason for this is that as the hydrogen and helium becomes more and more crushed under immense pressure in the giant gas planets, the physics of the behavior of matter enters a different regime. We see that there is a significant cluster of exoplanets with masses similar to the mass of Jupiter and with a size that is 10 to 20 times that of Earth (or between 1 to 2 times the size of Jupiter). However, we also see that there is a very interesting problem. Many of the giant exoplanets have a density that is even *less* than that of Saturn. But Saturn is already at around the minimum density you can get for something that is made predom-

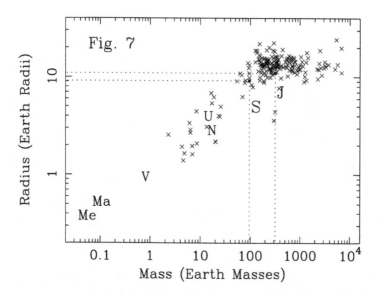

Figure 7: A diagram showing radius versus mass lower limit for the confirmed exoplanets in a subset of the October 2011 sample that have measurements of both radius and mass lower limit. Each exoplanet is marked with a cross. The radii and masses are expressed relative to the corresponding values for Earth (i.e., radii are shown in units of Earth radii, and masses are shown in units of Earth masses). Data are from the *Extrasolar Planets Encyclopedia*. For comparison, the radii and masses of the planets in our solar system are marked as *Me, V, E, Ma, J, S, U, N*, corresponding to Mercury, Venus, Earth, Mars, Jupiter, Saturn, Uranus, and Neptune, respectively (numerical values are from solarsystem.nasa.gov). Note that for clarity, the positions of Jupiter and Saturn correspond to the intersections of the dotted lines that are marked by the letters *J* and *S* respectively.

inantly of hydrogen and helium, and these are the least massive elements that exist. We shall see later that these giant exoplanets that are "underdense" are also very close to their host star (closer than Mercury is to our Sun), and they are known as hot Jupiters. A histogram of the density of all the exoplanets that have radii and mass lower limit measurements is shown in figure 8. Actually, the densities are lower limits because the masses are lower limits. The histogram very clearly shows a peak in density that is below the density of Saturn, the least dense planet in our solar system.

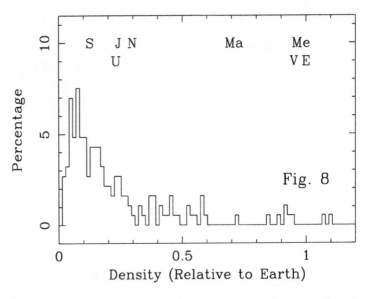

Figure 8: A histogram showing the distribution of the average density of the confirmed exoplanets in a subset of the October 2011 sample that have both measured radii and mass lower limits. Since the masses are lower limits, the densities are also lower limits. Note that a small fraction of exoplanets have densities that are greater than 1.2 times that of Earth and are not shown in this histogram in the interest of clarity (but all the densities are shown in figure 9). Plotted is the percentage of the subsample of exoplanets in each density interval. Densities are shown as ratios of the exoplanet densities to the density of the Earth (i.e., densities are shown in units of the density of the Earth). The densities are calculated from radii and mass lower limits, which are from the *Extrasolar Planets Encyclopedia*. For comparison, the densities of the planets in our solar system are marked as *Me, V, E, Ma, J, S, U, N*, corresponding to Mercury, Venus, Earth, Mars, Jupiter, Saturn, Uranus, and Neptune, respectively (numerical values of radii and masses used to calculate the densities are from solarsystem.nasa.gov).

We may ask if there are any patterns in the distribution of distance to the host star for the exoplanets that are in the low-density peak of the histogram in figure 8. The relation between the density and the distance from the host star is shown in figure 9, in units of Earth-Sun distance (i.e., in AU). Again, our solar system planets are shown on the plot for comparison.

What jumps out immediately from the plot is that the vast ma-

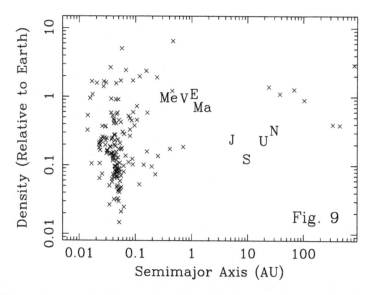

Figure 9: A diagram showing density versus semimajor axis (i.e., maximum planet-star separation) for the confirmed exoplanets in a subset of the October 2011 sample that have measurements of radius, mass lower limit, and semimajor axis. Each exoplanet is marked with a cross. The densities are expressed as ratios relative to the density of the Earth, and the semimajor axes are expressed as ratios relative to the Earth-Sun semimajor axis (i.e. they are in units of AU). Data are from the *Extrasolar Planets Encyclopedia*, and the densities are calculated values. For comparison, the planets in our solar system are marked as *Me, V, E, Ma, J, S, U, N*, corresponding to Mercury, Venus, Earth, Mars, Jupiter, Saturn, Uranus, and Neptune, respectively (numerical values are from solarsystem.nasa.gov). Amongst other things, the diagram shows that there is a significant clustering of exoplanets that have densities more than 10 times smaller than that of Earth, and star-planet separations more than 10 times smaller than that of Earth. These are the "underdense hot Jupiters," representing a kind of planet that does not have a counterpart in our solar system.

jority of the exoplanets are concentrated in a very narrow band as far as their distance from their host star is concerned. This band of occupancy is extremely close to the host star, in the range of about 1% to 10% of the Earth-Sun distance in our solar system. In particular, the unusually low-density, giant gas planets are very heavily concentrated in this narrow band, close to the host star. This proximity to the host star obviously means that the temperatures will be very high and this is of course why these close-in giant planets are known as hot Jupiters. The short distances from the host star also correspond to very short orbital periods (the length of year for an exoplanet) because objects in an orbit controlled by gravity have higher velocities for smaller star-planet separations (Kepler's third law of planetary motion). The clustering in the star-planet distances corresponds to orbital periods of about 2 to 6 Earth days, with some exoplanets having periods shorter than 2 days, going down to about 0.8 days. The reason for the clustering in such a narrow band is really not clear and is the subject of intensive research in the field. Could there be an observational bias (selection effect)? The majority of the planets with density measurements were discovered by the transit method, which does indeed favor finding close-in, short-period exoplanets. However, if selection effects were the only explanation, you would expect to see more of a gradual tapering off in the planet-star distance distribution, rather than the sudden cutoff at a period of about 6 days, or approximately 0.1 AU in star-planet distance. One possibility is that a planet at large distances (tens to thousands of AU) migrates so rapidly to small distances that the planet does not spend much of its life at intermediate distances, so the chances of finding a planet at intermediate distances is small. However, this still does not satisfactorily explain what causes the pileup at short distances from the host star, and it is not in fact understood what mechanisms stop a planet migrating too close to the host star. If rapid migration proceeded without a slowing down near the star, there would not be such a pronounced pileup. We can understand that no planets exist on the smaller side of around 0.01 AU from the host star because a planet that is too close to the host star will be subjected to a gravitational force that varies so much from point to point on the planet that it will be ripped apart (i.e., in jargon, *tidally disrupted*).

The closest distance of approach is known as the "Roche limit." The differential gravity creates forces that tend to distort and disrupt an object, and these are known as tidal forces. They lead to tidal disruption. A gaseous planet is generally held together only by its own gravity, but a rocky planet is held together by additional forces due to the electrostatic forces from atomic and molecular bonds. The latter type of planet may therefore venture closer to its host star than a gaseous planet before tidal disruption takes effect.

Physical models of the close-in exoplanets that have densities less than that of Saturn have a very hard time explaining the low density. To understand a possible resolution of this mystery we will need to take a detour concerning the nature of gravity.

Living Close to a Star

In this section we are going to do a thought experiment and imagine what it would be like to live on a planet that was orbiting very close to its host star, when the star-planet separation is only a few percent of the Earth-Sun separation.

However, we first need to look at the effects of gravity on the surface of a planet and how gravity determines orbital motion. One of the very first things we learn about as babies is falling. Everyone is familiar with Aristotle's realization that the acceleration of a falling object near the surface of the Earth is independent of the mass of the falling object. If the distance traveled by the falling object is not too great compared to the size of the Earth, then the acceleration is constant, making an object increase its speed of fall by the same amount for every equal interval of time. This happens to be about 21.8 miles per hour per second (or 35.3 kilometers per hour per second) near the surface of the Earth. We are of course neglecting the effects of air resistance in all of this discussion. In other words, when you fall, your speed relative to Earth will increase by 21.8 miles (35.3 kilometers) per hour *every second* so that it takes a mere 5 seconds of falling to reach a speed of approximately 100 miles per hour (or about 160 kilometers per hour). The actual value of the acceleration near the surface of a planet is determined by how much mass of the planet is packed into a given size, although it is not quite the same as density. Nevertheless, the acceleration is determined by the mass of the planet and the distance between the center of mass of the falling object and the center of mass of the planet. Therefore, on the surface of a planet, the acceleration, or surface gravity as it is otherwise known, is determined by the mass and radius of the planet.

The first thing to appreciate is that what you feel as your weight is the force of resistance to your tendency to fall. So when you are stationary on the ground and not falling, the back reaction of the Earth in response to you trying to fall into the ground is your weight. When you are in free fall and there is no ground beneath your feet to stop your free fall, there is no back reaction, and under these circumstances you are weightless. Everything in your body is falling at essentially the same rate so there is no back reaction between different parts of your body either. The second thing to appreciate is that an object that is in orbit *only* under the influence of gravity is also free falling. There is no fixed reference point in space if an object is in orbit around another, and they are really both falling towards each other. Thus the Moon is falling towards Earth and the Earth is falling towards the Moon. The reason why they don't crash into each other (i.e., why the Moon never hits the ground) is that they have a component of motion that is not along the line joining their centers, so they are in effect always trying to fall to hit each other but never do because their direction of motion is changing at every instant so they continuously miss each other. In the special case of a circular orbit, the physics works out such that the size and speed of the orbit take on values such that the amount that two objects fall towards each other is exactly compensated by their motion in a direction perpendicular to the fall so that the two objects remain at a fixed distance apart. Of course, if there are any energy losses (for example in the case of a satellite around Earth suffering from atmospheric drag), the two objects do get closer to each other, resulting in a decaying orbit, which ultimately does result in a crash. In general, orbits are elliptical, and the parameters of the ellipse depend on the initial conditions, such as how the two objects ended up in orbit, including the initial energy of the system, and the relative initial direction of motion. Whether an orbit is elliptical or circular, the objects partaking in the orbital motion are in free fall and therefore weightless.

Now let's consider the fact that Earth is also in orbit around the Sun. Aren't the Earth and the Sun falling towards each other as well? Yes, they are. Moreover, you are falling towards the Sun as well because you orbit the Sun as well. Associated with every acceleration there is a force, so what is the equivalent force that is pulling you towards the Sun? On the surface of the Earth the force from the ground that is stopping your fall is your weight. We can express the force between you and the Sun as some percentage of your regular Earth weight to get a sense of how small it is. The size of that force is a competition between the mass of the Sun (which is about a third of a million times that of Earth), and the distance between you and the Sun. It turns out that the force of the Sun pulling on something on the Earth's surface (e.g., you) is about 1,600 times smaller than the weight of that object on the surface of the Earth due to Earth's gravity. That is, about 0.06% of an object's weight is a good estimate of the force between that object and the Sun. For example, if you weigh 160 pounds (72.6

kilograms), you are being pulled towards the Sun by an equivalent weight of approximately 0.1 pounds (about 45 grams). You might think that is small, but it is not negligible: that is probably the weight of your ear! More importantly, if you were not in orbit around the Sun and were allowed to fall towards it head-on, you would reach a speed of 100 miles per hour (160 kilometers per hour) in just 2 hours and 2 minutes due to that "tiny" force of 1.6 ounces (45 grams).

There's more, because now we realize some interesting consequences. Suppose you are standing at a position and time on Earth such that the Sun is directly overhead and you weigh yourself and find that you weigh 160 pounds. The Earth is pulling you straight down but the Sun is pulling you straight up. Your net rate of falling (acceleration) towards Earth is slightly less because of the small opposing tendency to fall towards the Sun. Therefore your net weight is slightly less than it would be without the Sun. What happens when the Earth has spun half a revolution so that the Sun is on the other side of Earth? Well, you will want to fall towards the Earth and towards the Sun as before, but now these forces are in the *same direction* and not in opposition so the net result will be that you will weigh slightly more than you would without the Sun's gravity. So we have the curious situation that you would weigh less during the daytime and more during the night! To be sure, the spin of the Earth is associated with centripetal forces as well so the situation is much more complicated but the general point of this discussion is that the Sun's gravity can work both in opposition and in the same direction as an object's gravitational attraction towards its home planet. On Earth we don't notice this effect but imagine living on a planet that was so close to its host star that the star's gravitational effect at the surface of the planet was *comparable* to the planet's surface gravity due to its own mass. Then we get some very interesting effects. Let's ignore for the moment that you would not be able to survive on such a planet because it would be so hot that you would be evaporated. (You could imagine that your government has built heat-resistant, cocooned dwelling areas for you to live in.) In the morning, at "star-rise" you will feel yourself being pulled towards the horizon. The details depend on how your planet's spin axis is oriented, but for the sake of argument, let's suppose that you are on the equator and the planet's spin axis is perpendicular to its orbital plane. As well as wanting to fall towards the horizon, you will still want to fall to the center of your planet so your net weight will be in a diagonal direction into the ground. If you climbed onto a fence and jumped off, your "down" direction will be diagonal. As the morning wears on you feel the direction of "down" changing to become more and more horizontal. As the morning coffee break approaches you feel the urge to be dragged along the ground towards the horizon. The feeling passes and then you start to feel lighter and lighter as the star's gravitational pull becomes more and more "vertical." As lunchtime approaches and your star approaches the overhead position you continue to feel lighter, until finally, at noon, you could actually be weightless and float up into the air (depending on the relative strengths of gravity from your planet and your star, at your position). You can imagine how complicated the infrastructure of a city on such a planet would be in order to accommodate everyone, with everything starting to float around at lunchtime. Lunchtime cafes could be situated at the top of large pillars, for example (so that you could just "float" in). Parachutes could be available for people who have lingered too long after lunchtime. That's because after the star has passed overhead you start to feel heavier and heavier again as the planet's gravity once again dominates. At "star-set" you feel pulled towards the horizon, in the opposite direction that was "downwards" in the morning. At night your star is on the other side of your planet and as the night progresses you feel heavier and heavier. When the star is directly opposite you (with the planet in between you and the star), you feel incredibly weighed down and barely able to move. A stark contrast to the floating fun you were having at lunchtime.

What if your star's gravity were so strong that at high noon your star's gravity doesn't just balance the surface gravity on your planet, but is so strong that it dominates? Unfortunately, there would then be a possibility that you could be whisked off into space and end up escaping from your planet, possibly going into an eternal orbit around your star (again, ignoring evaporation). Presumably, your government will have built precautionary structures in heavily populated areas, but being caught in an unprotected area at lunchtime would carry a high risk. These are not frivolous observations. On the contrary they have direct relevance to the close-in exoplanets that are mysteriously underdense. You see, what if the planet is not made of rock but is gaseous or liquid? If the planet is so close to its host star that the starlit side of the planet experiences a gravitational force from the star that is comparable to, or larger than, the local surface gravity, what is going to happen? The planet is held together by gravity, and the gravity usually maintains the spherical shape because in equilibrium the inward collapse due to gravity is exactly balanced by the back reaction due to the compressibility limit. However, if the star's gravity in the part of the planet that is facing the star is so strong that it overcomes the local surface gravity, that material there will become weightless and may even become unbound from the planet. In other words, the planet could undergo continuous loss of mass if the material is gaseous or liquid (the liquid may of course vaporize in the process). Certainly, there will be a distortion in the shape of the planet, as the gravitational competition between the star and the planet will vary over different parts of the planet. The planet will be elongated along the line joining the planet and the star. The material on the dark side of the planet will experience reinforced acceleration in the same direction (towards the star) and that material will therefore be heavier, and the least likely to escape. The effects are similar to gravitational tides, but not quite. Tidal stress usually refers to the stress in the structure of the planet caused by the star's gravity having a different strength over different regions of the planet, causing it to be pulled apart and broken. Instead, what I have talked about here is simply the strength of the local surface gravity of the planet at a given point on the planet, compared with the acceleration towards the star at the same point. If the two are comparable, material will be simply lifted off from the surface layers of the planet. The process is dynamic because as the mass loss proceeds, the rate of loss will change because the surface gravity will change. The star-planet distance is also likely to change as the planet becomes less massive. The details will depend on the exact structure of the planet, amongst other things.

So, what about the new, close-in, anomalously underdense exoplanets that have been discovered? At the surface of these planets, which is stronger, the surface gravity due the planet itself, or the host star gravity? As it happens, one of the close-in hot Jupiters going by the name of HD 209458b (also known as *Osiris*) has become quite famous for claims that a plume, or comet-like tail, is present, indicating that the planet is evaporating, and is far-removed from a purely spherical equilibrium structure. The planet size is estimated to be about 40% larger than Jupiter and its mass lower limit is estimated to be about 70% of that of Jupiter. The tail is claimed to be as large as the planet itself. The surface of the planet indeed appears to be in the process of being torn away.

However, the modeling of the observational data is highly nontrivial, involving complex physics and many uncertainties. The evaporation scenario has been challenged, but the claim of mass escape prevails. There is at least agreement on the fact that the hydrogen atmosphere is highly extended. There is also agreement on the fact that if the atmosphere is escaping, then the observed and theoretical rate of mass loss is too small to affect evolution of the planet. However, this may not have been true in the past. It is important to remember that HD 209458b is too far away to be directly imaged. Everything is inferred from nonimaging spectroscopic and transit data. [3]

Now let's return to the general issue of the magnitude of the surface gravity of a close-in planet compared to the local planet-star gravitational force. Remember that for the Earth-Sun system the ratio of the surface gravity on Earth to the Sun's gravitational acceleration at the surface of the Earth is about 1, 600, which is high enough that there is no danger of the atmosphere being ripped off by the Sun, or for humans to float around in the air at lunchtime. In order to calculate the ratio of a planet's surface gravity to the free-fall acceleration towards the host star, we need four quantities. These are the masses of the planet and the star, the radius of the planet, and the planet-star distance. Strictly speaking, for the latter the relevant quantity is half of the length of the longest axis of the elliptical orbit, and is called the semimajor axis. It turns out that of the 192 exoplanets (in the October 2011 sample) that have radii measurements, 156 have all four measurements that we need. I have calculated a quantity that I will call the local planet-star acceleration ratio, for the 156 exoplanets and also for our own solar system planets, and figure 10 shows a histogram of the results. Actually what is calculated and shown is a lower limit on the surface acceleration ratio because the exoplanet masses are lower limits.

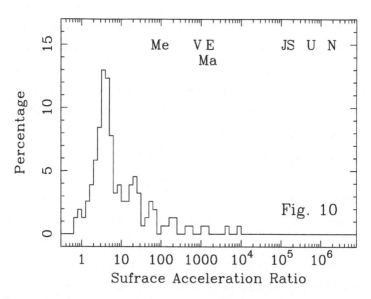

Figure 10: A histogram showing the distribution of a quantity called the "surface acceleration ratio" for a subset of the October 2011 sample of exoplanets for which the necessary measurements exist for the quantity to be calculated. Data are from the *Extrasolar Planets Encyclopedia*. The surface acceleration ratio is a measure of the tendency of the surface of the planet to be tugged away by the host star, due to the star-planet gravity becoming comparable to the planet's self-gravity at its surface. Mathematically, the ratio is simply the ratio of the local free-fall acceleration at the surface of the planet due to the planet, to the free-fall acceleration due to the host star. The histogram shows the percentage of exoplanets in the subset that lie in each surface acceleration ratio interval. The calculated surface acceleration ratios are actually lower limits because the exoplanet masses are lower limits. For comparison, the planets in our solar system are marked as *Me, V, E, Ma, J, S, U, N,* corresponding to Mercury, Venus, Earth, Mars, Jupiter, Saturn, Uranus, and Neptune respectively. The histogram shows that a substantial percentage of the exoplanets in the subset show a very strong clustering of the surface acceleration ratio of around 4. These are close-in hot Jupiters in which the surface layers of the planets are unstable, and they have no counterparts in our solar system.

Remarkably, there is a very strong peak at a value of about 4 for the acceleration ratio, and the histogram rapidly drops to zero at around a value of 1 for the ratio. In other words, a large fraction of the giant, close-in exoplanets are close to disruption due to the fact that they can't hold themselves together by gravity because the

101

host star's gravity is too strong and ripping off the surface layers. This is very intriguing. Why should so many exoplanets have a similar value of the ratio, and moreover, why should so many have a value that is so close to the critical "breakup" value of 1? This is a mystery. However, the fact that there are no exoplanets that have a value for the acceleration ratio much less than 1 makes sense because such exoplanets are prevented by the host star's gravity from ever holding themselves together.

In figure 11, I have plotted the local planet-star acceleration ratio (lower limits) against the orbital period (in Earth days) for the 156 exoplanets and also for our solar system planets. It can be seen that figure 11 clearly shows again the clustering of the giant exoplanets at around a value of the acceleration ratio of 4 and a range for the orbital period of around 1 to 4 days. The dotted line in figure 11 marks the value of 1 for the planet-star acceleration ratio, and we clearly see the "wall" that this represents, because the existence of exoplanets beyond this wall suddenly ceases.

Only three exoplanets are found below the wall, and these must be well on their way to breakup. We also see that all of our solar system planets lie well away from the breakup line, and the majority of the 156 exoplanets occupy a completely different regime of the diagram than the solar system planets. Another thing that we see in figure 11 upon closer inspection is that there seems to be a "hole" in the distribution close to the critical breakup line such that there is a deficit of exoplanets with a period of 2 to 3 days near the breakup regime. Instead, the distribution is split near the critical line, into two groups that have orbital periods of around 1 day and 4 days. This is puzzling behavior.

Now we see a possible solution for the existence of close-in, giant gaseous exoplanets that have an apparent inexplicably low density. These underdense exoplanets with masses between 1 to

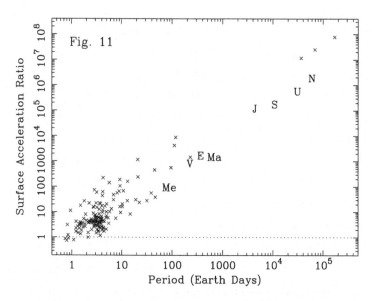

Figure 11: A diagram showing a quantity called the "surface acceleration ratio" versus orbital period (in Earth days) for a subset of the October 2011 sample of exoplanets for which an orbital period has been measured, and for which the necessary measurements exist for the surface acceleration ratio to be calculated. Data are from the *Extrasolar Planets Encyclopedia*. The surface acceleration ratio is a measure of the tendency of the surface of the planet to be tugged away by the host star, due to the star-planet gravity becoming comparable to the planet's self-gravity at its surface (see figure 10). The calculated values of the surface acceleration ratios are actually lower limits because the exoplanet masses are lower limits. For comparison, the planets in our solar system are marked as *Me*, *V*, *E*, *Ma*, *J*, *S*, *U*, *N*, corresponding to Mercury, Venus, Earth, Mars, Jupiter, Saturn, Uranus, and Neptune respectively. The diagram shows that a substantial percentage of the exoplanets in the subset show a very strong clustering of the surface acceleration ratio of around 4 and orbital periods of around 2 to 4 Earth days. These are close-in, short-period, hot Jupiters in which the surface layers of the planets are unstable, and they have no counterparts in our solar system. There appears to be a cutoff (or "wall", corresponding to the dotted line) at a value of the surface acceleration ratio of around 1, below which no exoplanets are found. This likely corresponds to a regime in which exoplanets cannot exist due to the star-planet gravity overwhelming a planet's self-gravity.

2 Jupiters and sizes within a factor of 2 of the size of Jupiter are precisely those exoplanets that cannot hold themselves together by their own gravity because of their proximity to their host star.

These underdense planets are just those that have a local planet-star acceleration ratio (on the star-facing side) that is about 4 or less. They are losing mass from the surface layers because of the domination of the host star's gravity over the local surface gravity on the planet, on the star-facing side. Calculations in the current scientific literature typically show radius versus mass diagrams, and for various chemical compositions and model assumptions, show lines of constant density overlaid on the exoplanet data points. The fact that the models don't explain the very low-density, close-in exoplanets is then referred to as the "radius anomaly" in the scientific literature (i.e., the radius is too large for the given mass). However, such model calculations assume some sort of equilibrium (i.e., a stable configuration) but such an assumption is clearly not valid for a system in which the surface layers of the planet are being ripped off by the host star. It is not surprising that equilibrium models cannot account for the low density. The planet is going to be in considerable turmoil, and parts of the planet may even be "patchy," resulting in a lower average density. This has finally been pointed out in recent studies, but a recent review still states, "...no mechanism has gained the consensus of the community as being clearly important," where "mechanism" here refers to a mechanism for explaining the "radius anomaly." [4]

I emphasize that the local planet-star acceleration ratio is a subtly different indicator to the standard measure of gravitational tides that also distort an orbiting object and can also break it up. This will be easier to visualize with a specific example, so let's take the case of a planet and a star in orbit due to their mutual gravity. Let's suppose that the planet is held together as a coherent body only by its own gravity (this is not necessarily true for a rocky planet but is more likely to be the case for a gaseous or fluid planet). Gravitational tides refer to a situation in which the star-planet gravitational force on the star-facing side of the planet is significantly different to that on the side of the planet facing away from the star. The gravitational force keeping the planet in orbit varies all over the planet in fact. In other words, on the one hand the planet is trying to free fall as a coherent single unit, but the differential forces have the effect of trying to make different parts of the body free fall at different rates. The resulting stress and tension can break up the planet and this can be quantified by what is known as the *Roche limit*, the critical distance between the planet and the star that would result in tidal disruption if such a close planet-star proximity were to be attained. In other words, the planet has to stay outside of the Roche distance to avoid destruction.

In the case of ocean tides on Earth that are caused by the proximity of the Moon, the rate of free fall of the Earth towards the Moon is different for the side of the Earth closest to the Moon than it is for the opposite side of Earth. Since the Earth moves as one coherent body, with a single free-fall rate, the resulting tension causes distortion of the shape of the oceans. In more extreme cases the tension can be so great as to break up even a rocky body. These tidal effects are the ones that are usually discussed and modeled in the context of close-in exoplanets. However, the local planet-star acceleration ratio measures something that is manifestly different. It measures how well the surface layers of a planet can remain bound to the planet by its own gravity, as opposed to being accelerated towards the host star. Thus, the planet-star acceleration ratio is more relevant for gaseous planets, and the atmospheres of rocky planets.

Obviously, the tidal breakup criterion and the surface mass loss criterion (i.e., the planet-star acceleration ratio) are closely related and we can work out the relationship between them. It turns out that the surface mass loss condition is always reached first (before tidal breakup), but the two conditions become more and more similar the larger the physical size of the planet. What I mean here by the size of the planet is the radius of the planet compared to the size of the orbit. So, for planets that are small, or large planets that are far away for the star, the surface mass loss criterion is reached well before tidal breakup. However, for planets that are large compared to the size of the orbit (for example, the close-in giant exoplanets), the surface mass loss and tidal breakup become important at around the same time, but still, surface mass loss begins before breakup. What is interesting is that if a planet is losing mass and the size of the planet is also changing as it loses mass, the critical conditions for both surface mass loss and tidal breakup change. So, the question is, do these changes make the planet less stable or more stable? The answer really depends on the details of the mass loss and how the size changes, so more information is required. However, it is likely that a decrease in size and mass actually leads to more stability because the remaining surface layers become more strongly bound to the planet, and the differential tidal forces across the planet become less. So it is possible that the process is self-terminating. In other words, a planet that is in the zone of prolific mass loss and instability against tidal disruption could end up reaching a more compact stable configuration after shedding enough mass. This is interesting because it is a possible scenario for making a small rocky planet out of a large gas giant (if gas giants have rocky cores).

A direct comparison of the surface mass loss criterion and the tidal breakup criterion is shown in figure 12. Here I have shown the 156 planets for which all the necessary measurements are available to calculate both criteria for each planet (in the October 2011 sample). Note that the tidal disruption parameter values on the vertical axis of figure 12 are actually upper limits because the exoplanet masses are lower limits. In other words an exoplanet could be more stable against tidal disruption than the calculated upper limit indicates. Our solar system planets are also shown in figure 12 for comparison.

In the plot, points that are further away horizontally from the value of 1 on the horizontal axis are more and more stable against surface mass loss. Points that are further away vertically from the value of 1 are more and more stable against tidal breakup. Therefore exoplanets occupying the top left-hand part of the plot are the most unstable against both surface mass loss and tidal breakup. As expected, the group of close-in giant gaseous planets occupies the top left-hand corner of the plot. It is remarkable that so many exoplanets should strongly cluster around the same values of the critical quantities. It begs an explanation of why so many of these exoplanets are so similar in that they are all so close to being tidally disrupted, as well as having a planet-star acceleration ratio that is near the critical value (resulting in significant mass loss). It remains a mystery. However, I have a "crazy" suggestion, but it will have to wait until after we have looked at planet formation.

Orbital Period and Size

So far we have just looked at the subset of about a third of the exoplanets in the October 2011 sample (i.e. those exoplanets that have measurements of their radius). Now we will look at the larger population. Of the 692 exoplanets, 686 of them have some kind of mass lower limit (except for four of these, which have mass *upper* limits, because they reside in multiplanet systems). Of the 692 exoplanets, 669 have orbital period measurements, and 654 have measurements of the size of the orbit (the semimajor axis is the size indicator, which is half of the long axis of an elliptical orbit).

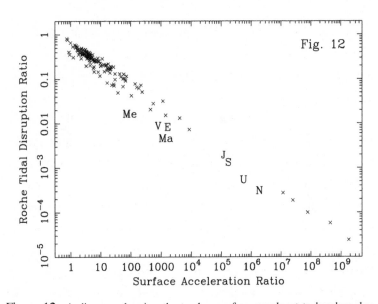

Figure 12: A diagram showing the tendency of an exoplanet to breakup due to tidal disruption (by its host star), versus the tendency to lose mass from its surface layers. Exoplanets are more likely to be tidally disrupted the higher they are on the diagram in the vertical direction, and they are more likely to lose mass from their surface layers the further to the left they are on the diagram. Therefore exoplanets located at the top left of the diagram are the most unstable and those located at the bottom right of the diagram are the most stable. What is actually plotted on the vertical axis is the ratio of a measure of the "Roche limit distance" (see text) from the host star, to the maximum orbital radius. This ratio approaches 1 as the orbital size approaches the critical threshold distance from the host star for tidal disruption to set in. The calculated values of the tidal disruption parameters are actually upper limits because the exoplanet masses are lower limits. On the horizontal axis a quantity called the "surface acceleration ratio," is plotted, and it is a measure of the tendency of the surface of the planet to be tugged away by the host star, due to the star-planet gravity becoming comparable to the planet's self-gravity at its surface (see figure 10). The calculated values of the surface acceleration ratio are actually lower limits because the exoplanet masses are lower limits. The diagram is constructed from a subset of the October 2011 sample of exoplanets for which the necessary measurements exist for the relevant quantities to be calculated. Data are from the *Extrasolar Planets Encyclopedia*. For comparison, the planets in our solar system are marked as *Me, V, E, Ma, J, S, U, N*, corresponding to Mercury, Venus, Earth, Mars, Jupiter, Saturn, Uranus, and Neptune respectively. As expected, the close-in hot Jupiters occupy the top left (most unstable region) of the diagram, and they have no counterparts in our solar system.

The period of a planet is the time it takes to make one complete revolution around its star, so it is essentially the planet's "year" and we will refer to it in units of Earth days. The period of the orbit is related to the size of the orbit by physics (Newton's and Kepler's laws of motion under gravity) but additional information is needed in order to actually extract the semimajor axis from the period. The mass of the star and the eccentricity of the orbit need to be known.

A histogram of orbital period (in Earth days) is shown in figure 13, which also shows the positions of our solar system planets and Pluto for comparison. A very interesting result is immediately apparent: the distribution has two clear peaks (i.e., it is bimodal).

The first peak, between about 2 to 4 Earth days, corresponds to the population of close-in, gas giants (or hot Jupiters) that I have talked about extensively already. Then there is a relative deficit of exoplanets until we get to about 100 Earth days. From there to about 6,000 Earth days is the second peak in the distribution. This second peak is much broader than the one for the close-in giant exoplanets. Since the size of the orbit is closely related to the orbital period, we expect the distribution of orbital size to also be double-peaked, and indeed it is, and this is shown in figure 14. The units are AU (the nominal Earth-Sun distance), and our solar system planets and Pluto are again shown for comparison.

As we have seen earlier, the short-period peak corresponds to about 0.02 to 0.1 AU for the close-in gas giants, and the second peak corresponds to a range of about 0.8 to 6 AU. The deficit is therefore in the approximate range 0.1 to 0.8 AU.

The bimodality in the orbital period and orbital size distributions has been known since 2003. [5] However, there is no satisfactory explanation for it in the scientific literature. It is hard for a selection effect to account for the data. If there were gas gi-

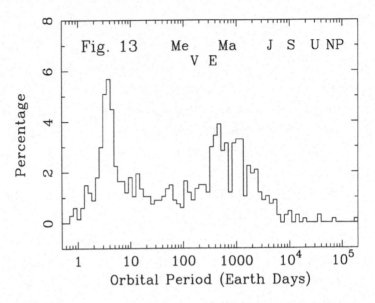

Figure 13: A histogram showing the distribution of the orbital periods of the confirmed exoplanets in a subset of the October 2011 sample that have measured orbital periods. Plotted is the percentage of the subsample of exoplanets in each period interval. The exoplanet periods are shown in units of Earth days. The histogram shows a bimodal distribution in orbital period. Data are from the *Extrasolar Planets Encyclopedia*. For comparison, the periods of objects in our solar system are marked as *Me, V, E, Ma, J, S, U, N, P* corresponding to Mercury, Venus, Earth, Mars, Jupiter, Saturn, Uranus, Neptune, and Pluto respectively (numerical values are from solarsystem.nasa.gov).

ants in the deficit region with orbital periods of 10 to 100 days, such planets would have been found by the transit or velocimetry methods. Both methods are biased towards smaller orbits (shorter periods) and large masses, but the fact that there are plenty of planets on either side of the deficit region means that selection effects alone cannot explain the bimodality. A more subtle selection effect could be at work if the deficit region truly contains a different type of planet compared to the rest of the distribution. In other words, if a large proportion of exoplanets in the deficit region really were small, low-mass planets such as the terrestrial planets in our solar

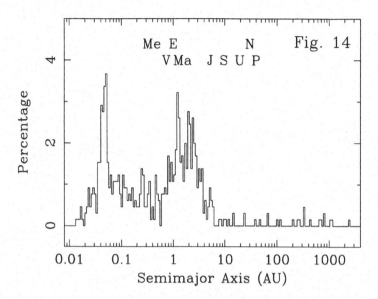

Figure 14: A histogram showing the distribution of the orbital size (semimajor axis) of the confirmed exoplanets in a subset of the October 2011 sample that have measured semimajor axis values. Plotted is the percentage of the subsample of exoplanets in each interval of semimajor axis. The exoplanet semimajor axes are shown in units of AU (i.e., in units of the semimajor axis of the Earth's orbit around the Sun). The histogram shows a bimodal distribution in semimajor axis (as expected from the bimodal distribution in orbital period shown in figure 13). Data are from the *Extrasolar Planets Encyclopedia*. For comparison, the semimajor axes of objects in our solar system are marked as *Me, V, E, Ma, J, S, U, N, P*, corresponding to Mercury, Venus, Earth, Mars, Jupiter, Saturn, Uranus, Neptune, and Pluto respectively (numerical values are from solarsystem.nasa.gov).

system, they would be missed by observational bias.

Scenarios that attempt to explain the orbital period and size deficit by outer planets migrating inwards have a hard time explaining the pileup of hot Jupiters close to the host star. In these scenarios the deficit arises if the migration is such that a relatively little proportion of time is spent by the planet while migrating in the deficit region so that the chances of finding a planet there are reduced. However, by the same token, the planet pileup close to the host star means that the planets that have migrated so close to

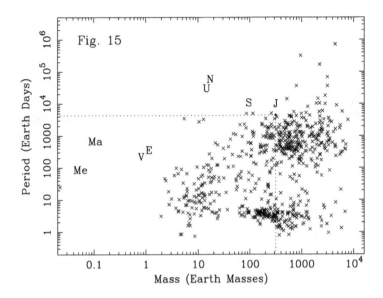

Figure 15: A diagram showing the orbital period (in units of Earth days) versus exoplanet mass limit (in units of Earth masses) for a subset of the October 2011 sample of exoplanets that have orbital period and mass estimates. Data are from the *Extrasolar Planets Encyclopedia*. For comparison, the planets in our solar system are marked as *Me, V, E, Ma, J, S, U, N*, corresponding to Mercury, Venus, Earth, Mars, Jupiter, Saturn, Uranus, and Neptune respectively (numerical values are from solarsystem.nasa.gov). The short-period exoplanets appear to have a wider range in the mass parameter than the long-period exoplanets. See text for further interpretation of the diagram. Note that for clarity, the position of Jupiter corresponds to the intersection of the dotted lines that is marked with the letter *J*.

the host star spend a relatively long time in that phase. This has not been explained. [6]

We can get an additional perspective into the orbital period distribution if we plot orbital period versus mass of the exoplanets. This is shown in figure 15. The mass (lower or upper limit) is plotted in Earth masses and the period is plotted in days. Our solar system planets are shown for direct comparison.

The two clusters of exoplanets can be seen as a higher concentration of points in the two patches around 2 to 4 days in orbital period and around 100 to 6,000 days in orbital period. However what the plot reveals is that all of the planets in the concentrated patches are giants, with a mass greater than 100 Earth masses or so. We see that actually, for the intermediate orbital period range of about 10 to 100 days, there is actually no preference in orbital period and planets of all masses, from 2 to 10,000 Earth masses are found there. We can put this another way and say that the double-peaked orbital period distribution reflects the fact that giant exoplanets with a mass greater than 100 Earth masses are preferentially, but not exclusively, found to reside very close to or very far from the host star, but there is no such preference for exoplanets that are less massive than about 100 Earth masses. However, we should exercise some caution in interpretation because the sample could still be statistically incomplete in the region of intermediate orbital periods. Certainly, the lack of exoplanets to the left (low masses) and to the top (very long orbital periods of more than 10 years) is purely an observational selection effect because of the difficulty of finding and confirming planets with low masses and/or very long periods.

Another interesting aspect of figure 15 is the apparent lack of exoplanets that have an orbital period of less than 3 days and a mass less than 1 Jupiter (about 318 Earth masses). There are a few exoplanets with a period of less than 3 days that have a mass less than 10 Earth masses but there is a remarkable absence of masses in the range 10 to 300 Earth masses. This is another puzzle that demands an explanation. Models can be tweaked to explain the high-mass side of the effect but then they fail to explain the reappearance of exoplanets with periods of less than 3 days and masses less than 10 Earth masses. [7]

Yet one more thing that figure 15 shows is that the terrestrial solar system planets lie in a part of the diagram that is not occupied by an overwhelming majority of exoplanets. Even the outer planets in our solar system are at the edge of the region in the diagram occupied by exoplanets. This emphasizes the fact that there is still a way to go to find large numbers of so-called "Earthlike" planets. Note that the very low-mass planet at the extreme left of the diagram in figure 15 (which has only about 2% of the mass of the Earth) is one the planets found around a pulsar (a compact spinning remnant of a dead star). Aside from this "odd" planet, the lowest mass exoplanet in the diagram has a mass of about 2 Earth masses, and a period (i.e., "year") of only 3 Earth days, so it is extremely close to its host star (about 0.04 AU). Unfortunately, the radius of this exoplanet, known as Gliese 581e, has not been measured, so many properties such as density and composition can only be guessed until the size of the exoplanet is measured.

Finally, figure 15 shows some evidence of a third, but weaker, clustering of exoplanets occupying the region between about 2 to 20 Earth masses and about 2 to 20 days in orbital period. However, it is too early to eliminate an artificial origin of this clustering.

Exoplanet Eccentricities

Orbits are elliptical in shape because of the mathematical forms of Newton's laws of gravity and motion. Kepler's laws of planetary motion are simply applications of Newton's laws, which are very general. The eccentricity of an ellipse is a measure of how "squashed" the ellipse is compared to a circle, and has a value in the range 0 to 1. An eccentricity of 0 is a special case of an ellipse that is a circle. Higher and higher values of the eccentricity characterize more and more elongated ellipses. The eccentricity of the orbit of a planet is affected by several factors. The initial conditions during the formation of the planet will affect the eccentricity in complex ways. Impacts and collisions with other bodies can

114

also affect the eccentricity. Also, gravitational tides, which cause stress and tension throughout the planet and affect its spin and orbital rotation, can cause an elliptical orbit to eventually become circular if the planet is very close to the host star.

All current planet-formation scenarios involve some sort of rotating disk around the host star, from which planets are formed (I will talk about this more later). The circular motion of the disk should give rise to planets that have zero or small eccentricities. What is actually found from the exoplanet sample of October 2011 is that a significant fraction of the exoplanets has an eccentricity that is not zero. The number of exoplanets in the October 2011 sample with an eccentricity greater than zero is 497. Of these, 361 (about 52%) have an eccentricity greater than 0.1, and 62 (about 9%) have an eccentricity greater than 0.5. Thus, there is a drop in numbers for eccentricities larger than 0.5. [8] There is a tentative suggestion that the median eccentricity for giant planets is higher than it is for lower mass planets. However, interpretation of the data requires caution. It is true that there are many more low-mass exoplanets that have eccentricities lower than 0.5 than there are low-mass exoplanets with eccentricities higher than 0.5, but the latter may be unfavorable for detection. Highly eccentric orbits require a higher frequency of "snapshots." On the other hand, there is a definite lack of giant planets that have a mass greater than about 3 Jupiter masses (or about $1,000$ Earth masses) and circular orbits. More specifically, a diagram showing eccentricity versus planet mass is essentially unpopulated above 3 Jupiter masses and below an eccentricity of 0.1. This cannot be explained by observational bias because planets with such large masses and circular orbits should actually be easier to detect than planets with small masses and large eccentricities.

In contrast to the population of exoplanets, most of our solar system planets have nearly circular orbits. Seven of the eight solar system planets have eccentricities less than 0.1, and the eccentricity for Mercury is about 0.20.

Is Our Solar System Typical?

It should be clear from the preceding discussions that our solar system is far from typical, at least in comparison with the exoplanets that have been discovered so far. Many of the new findings are still hailed in the scientific literature as surprising and unexpected, in particular the preponderance of the hot Jupiters. Such reactions are puzzling because ideas and theories about the properties of planets and about planet formation were based on a sample of one solar system. Before the discovery of exoplanets, models of planets and the solar system were tweaked and nudged to explain only the solar system. This is understandable to the extent that no other data existed, but then the expression of surprise is not understandable because the notion that all other solar systems would be expected to be like ours was an assumption that had no factual basis. Even an untrained scientist knows very well that you cannot make general conclusions based on a sample of one. Our solar system is not a natural result of the models that were used to explain it because there are many variables, approximations, and input assumptions about what physical interactions are important to include, and about the physical conditions that prevail in circumstances for which there is no information. These unknowns required various assumptions and the assumptions were selected and adjusted to reproduce the properties of the one and only system that we knew about. Even with such selective and directed modeling, not all of the properties of our solar system could be explained (more about this later).

There is no counterpart in our solar system to the hot Jupiters, those gas giants that orbit closer to their host star than 0.1 AU. Yet the fact that a large proportion of confirmed exoplanets are hot Jupiters seems to indicate that this variety of planet may actually be very common. There also large numbers of ice giants located between about 1 to 5 AU from the host star, with masses ranging from about 0.5 to 30 or so Jupiter masses. This range of mass can actually be found at all distances from the host star, but the two groups located in the approximate ranges 0.01 to 0.04 AU and 1 to 5 AU contain the majority of the giant planets.

The regime of exoplanet properties between approximately 2 Earth masses and 100 Earth masses (the latter being about a third of a Jupiter mass), and a star-planet distance in the range 0.01 to 5 AU is populated by exoplanets that also have no counterpart in our solar system. This is the regime of the super-Earths and the hot and cold Neptunes (our Neptune is cold but further than 5 AU). There are no counterparts to the terrestrial planets in our solar system in the October 2011 sample, although the few super-Earths in the lower mass range begin to approach that regime. For the latest information check the first two columns of numbers in the *Extrasolar Planets Encyclopedia* catalog (you can sort by mass or radius by clicking on the respective column headings).

Planet Formation and Migration

The current thinking about how planets are made from scratch revolves around the idea that the host star, in its early life, is associated with a flattened, rotating disk-like structure. The disk is made of the same stuff as the star, mostly hydrogen and helium, plus heavier elements such as carbon, nitrogen, oxygen, neon, silicon, iron, nickel and more. The elements heavier than hydrogen and helium are collectively referred to as metals in the trade, even though some of them are not metals in the traditional sense of the word. The elements that have high condensation temperatures are also referred to as refractory elements (as opposed to "volatiles"). Although the percentage of metals is tiny (in total less than 1%), it is plenty to make planets. However, the total mass required of a protoplanetary disk is uncertain but believed to be a minimum of 1 to 2% of the mass of the host star. The disk also contains compounds of various kinds and some of these are referred to by the highly technical term, "dust." Dust can have various compositions involving some combination of the metals (for example, oxygen and iron). A common property of dust is that it reprocesses the radiation from the host star into infrared radiation and this is observable by infrared telescopes (the disk is dark in visible light because it absorbs visible light). Compounds that have been found in meteorites are thought to be direct relics of the ancient disk that was once associated with our Sun. The disk is often referred to as a "debris disk" because of the various coagulations of compounds in the disk.

Another property of the debris disks that is often touted as favoring association with planet formation is that the material in the disk rotates because the star from which it is formed spins. If planets are formed form the disk they will end up orbiting the star in the same direction as the star's rotation. However, this has now become problematic because the number of exoplanets with spin-orbit misalignment, and with outright retrograde motion, has become significant. Exoplanets that rotate in the opposite direction to the star's spin are then difficult to explain. A recent study concluded that *most* hot Jupiters have a spin-orbit misalignment, with as many as 25% actually showing retrograde motion. Solutions have generally focused around perturbations by unseen massive objects such as other planets or a distant star that is a member of a binary pair. There is also something known as the "Kozai effect" which is basically an "interaction" between the orbital tilt and the eccentricity of the orbit, in the sense that the latter two quantities can vary such that the orbit is able to exchange one in favor over the other, in a kind of to-and-fro "resonance." The origin of the distribution of the eccentricity is also problematic for standard planet-formation scenarios and proposed explanations again involve planet-planet interactions after the planet-formation phase. However, the solutions to the problems presented by skewed and/or highly elliptical orbits remain speculative because the objects responsible for the perturbations have yet to be found in a single case. [9]

Dusty debris disks have actually been observed, but they are only found in stars that are very young, less than 10 million years old or so (compared to a typical lifetime of 10 billion years for a sunlike star). Recently a particular exoplanet was directly imaged and found to be residing in a debris disk. This is the exoplanet known as Fomalhaut b, which is a gas giant at a large distance from its host star. However, observing a planet in a debris disk does not prove that it was formed in the disk, and it does not indicate the mechanism of formation. A planet that is *unambiguously* in the process of formation has never been observed. In other words, there is no direct proof that planets form in the disk, it is an inference made simply from the coexistence of the planet and the disk. If you catch someone at a crime scene, you have to establish many things to prove that the person committed the crime. The same applies to a more recent case (Kraus and Ireland 2011, http://arxiv.org/pdf/1110.3808v1, *LkCa 15: A Young Planet Caught at Formation?*). The question mark at the end of the title was conveniently omitted in a flurry of popular news stories.

The paper discussed caveats pertaining to the fact the "blob" could just be a "random" chunk of matter that has nothing to do with planet formation. The authors themselves stated that further observations should be made to make conclusive deductions, but the spin in popular news stories (even those on dedicated astronomy websites) did not reflect this bottom line.

Current planet-formation scenarios all require the production of large chunks of matter, and at least one chunk that is large enough for it to start attracting other chunks by its own gravity. These chunks can be formed either from the "bottom up" or from the "top down". The former process starts off with the coagulation of tiny dust grains and the latter starts by gravitational instabilities in the debris disk (which is gaseous and turbulent). The bottom-up approach, known as "core accretion" is the one that is favored in the current literature. Accretion refers to the collection and aggregation of matter onto the object in question and the bottom-up scenario is known as (the) "core accretion" (theory). However, as we will see, neither approach actually works. In the core accretion scenario the chunks of matter have to reach a minimum threshold mass for gravity to kick in enough for the chunk to start growing to become what is known as a planetesimal. However, that threshold is huge. Planetesimals have to have a size of hundreds to thousands of kilometers before their mutual gravity becomes strong enough to start growing into a planet. While it is easy to get meteorite-sized objects from the coagulation of tiny dust grains and the like, once you get to boulders that are around a meter large in size, you hit a showstopper. As you know, boulders do not stick together with their own gravity. What's worse is that these boulders are theorized to be in a highly turbulent environment, traveling at very high speeds relative to each other. When these boulders collide, again as you would expect from common sense, instead of sticking together, they smash each other into smaller pieces. That's the exact opposite of what you need to build a planet. It is a show-stopper. Nobody has been able to think of a way for boulders to stick together at all. As a scientist, I find what happens next deeply disturbing. Researchers admit that the problem is severe but then they simply assume that somehow boulders did manage to stick together and then they proceed to develop a sophisticated theory of planet formation assuming that the critical mass was formed. There is no known mechanism to make boulders or kilometer-sized objects stick together and prevent them from smashing each other apart so it is almost like invoking some kind of magic. It is certainly true that if there does exist a large enough core, that core will accrete material and grow larger by means of gravity. However, if the foundation of planet-formation theory is an impossibility, why should one have any reason to believe that the theory on top of that foundation is a true description of what actually takes place? If a completely different approach is required, how do we know whether that alternative mechanism continues to operate (or not) during gravitational growth? To use an analogy, the current theory of planet formation is like making a detailed budget for how you are going to spend a billion dollars when you only have a thousand dollars in the bank. It is of course possible that could run into a billion dollars tomorrow. However, shouldn't you be spending more time figuring out how to acquire a billion dollars, rather than how you are going to spend it? If you don't acquire the billion dollars, your budget will be irrelevant and wrong.

Just when you think it couldn't get any worse, it does. You see, planet-formation models are extremely complex and there are many unknowns. You have to decide what your initial conditions are, what the physical state at every point in the protoplanetary disk is, what approximations and assumptions you are going to make, and so on and so forth. There is a lot that is put in by hand. For example, in some simulations of terrestrial planet formation, Jupiter and Saturn are simply put in place by hand, with the assumption that they were *somehow already formed.* [10] This is because Jupiter's gravity is so strong that it affects the aggregation of mass onto a smaller object that might be very far away. In other words, Jupiter's gravity would have affected the formation of Earth. If you don't put in Jupiter by hand, things can go crazy and you might not get the answers that you want. There are many other knobs and dials that can be tweaked by hand in a planet-formation model. Obviously the more there is to tweak, the less useful the model is in giving insight into the true planet-formation mechanisms. The goal for any scientific model should be to have as few adjustable parameters as possible in order to reproduce the observables in the "real world." Prior to the discovery of exoplanets the statistical predictions of the standard planet-formation scenarios for samples of exoplanets could not be tested. Now they *can* be tested and the results are already a disaster. Again, I am puzzled by expressions of surprise in the scientific literature, that the predictions of the standard model conflict with certain observational facts. Why should it be surprising? The theory is based on a hypothesis for which there is no known physical mechanism. It was only a question of time before a conflict between theory and observation was exposed, and indeed this was concisely illustrated by a paper published in a peer-reviewed journal in 2010. [11] This paper actually reveals one of the most fundamental problems with the current core accretion scenario *in addition* to the "no sticking" problem. The paper examines the properties of a sample of exoplanets that have an orbital period of less than 50 days and addresses how the number of exoplanets found in various mass ranges compares with the prediction of the core accretion scenario.

The title of the paper is "The Occurrence and Mass Distribution of Close-in super-Earths, Neptunes, and Jupiters." The authors studied exoplanets around 166 stars and found that, compared to the theoretical prediction, there is a statistically significant overdensity of planets in the mass range of 5 to 30 Earth masses (with orbital periods less than 50 days). The core accretion scenario actually predicts a "planet desert" in this regime, but no such deficit of planets is found. The severe conflict between the prediction and reality is heavily underplayed. The abstract states, "This region of the parameter space is in fact well populated, implying such models need substantial revision." In other words, the paper does not state that the current models could be completely wrong, or that the current models are based on an impossible premise, requiring the need for a paradigm shift. The paper never really clarifies what is meant by "substantial revision" and stops at that. The observational result reported in the paper is robust to selection effects because the "completeness correction" that was applied to account for planets that might have been missed by observational bias *increases* the number of planets in a given mass range. In other words any observational bias would work in the direction of reducing the disagreement between predicted and observed planet numbers, not increasing it. The paper does not of course mention anywhere that the models that are referred to require the sticking together of boulders by hand.

Now, the reason why the core accretion scenario predicts a planet desert for the mass range of around 5 to 30 Earth masses and an orbital period of less than 50 days is very simple. A growing planet in this regime undergoes rapid evolution in only one of two ways. Either interactions with the environment result in energy loss that makes the planet rapidly spiral inwards towards its host star, or rapid runaway accretion of gaseous material grows the planet too quickly into a giant. Since planets with masses of 5 to 30 Earth masses are short-lived in the model, a snapshot survey should not find many of them. It is quite remarkable that even with a foundation that forces the sticking together of boulder-sized and mountain-sized chunks by hand, even with all the additional knobs and dials that are available to tweak, the standard model of planet formation fails in a very fundamental way. The spiraling into the host star and the runaway growth scenarios are both controlled by basic physics, and neither scenario can be made to go away by the knobs and dials in the model. This should be a big clue that something is very wrong with the current paradigm, rather than indicating that there is a problem with the details of the model. What is interesting is that both problems of planet migration and runaway accretion were known before, but the problems were never confronted with real data until recently.

Unfortunately, school and college textbooks on science and astronomy give an extremely sanitized handwaving account of planet formation and persuade the reader that either planet formation is quite well understood, or at worst that there are some small niggling loose ends to tie up. How do research papers approach the awkward and embarrassing "no sticking" problem? There are a number of ways, but allow me to give some specific examples from a recent review paper on the subject of planet formation. [12] In the introduction of the paper, the author states that, "The planetesimal hypothesis is widely accepted today as the basis of terrestrial planet formation." No citation is given to back up the use of the term "widely accepted." I don't know what fraction of scientists accept the hypothesis. If the wide acceptance is true then it is deeply troubling because it means that a large number of scientists are willing to accept a hypothesis that is currently known to be physically impossible, and their acceptance reflects the "belief" that some day someone will find a way to make boulders and mountains that are colliding at high speeds stick together instead of being smashed apart. If the "wide acceptance" statement is not true, then it is also troubling because then the implication is that the author has made a statement about the hypothesis that is not supported by fact. In any case, most of the paper discusses what happens or could happen *assuming* that planetesimals can, and have formed (by an unspecified mechanism). Towards the end of the paper the author does address the "no sticking" problem directly. The author states that, "In the particle-sticking model, growth is especially difficult for boulder-sized bodies, a problem referred to as the 'meter-size barrier.' " The author explains that these boulders could have relative speeds of several tens of miles per hour and goes on to say that, "These collisions probably lead to erosion rather than growth." The way the problem is presented here is typical: the severity of the problem is first underplayed, and a name is invented for the problem so that one can put it aside and continue as if the problem didn't exist, on the premise that somebody will solve it. The assumption here is of course that it *is* solvable, but there is no basis for that assumption. The term "especially difficult" is used instead of "impossible" or "unexplained," and the word "probably" is used to soften the fact that collisions *will* hinder growth. The wording implies that there may be some circumstances under which boulder-sized bodies *could* stick together, and some circumstances under which collisions *could* lead to growth. However, no citations to any study are given that demonstrate either of these two things. The wording reflects "hope" only, and that is not science. The use of the term "meter-sized barrier" implies that the barrier is in some way a property of physics, or physical conditions in the model. Where is this barrier? There *is* no physical "barrier." Really, the only barrier is in the human mind being unable to let go of a paradigm that doesn't work and come up with a new paradigm. There are only two possibilities. Either there is a regime of physical conditions that humans have not thought of that would allow meter-sized objects to aggregate into planetesimals, or the core-accretion, bottom-up scenario is completely wrong. In either case, the "barrier" is in the human mind and has nothing to do with physics. The review paper does not mention any need for a paradigm shift, and instead states that, "In the light of these problems, the hunt is on for a new mechanism for planetesimal formation that can operate in a turbulent environment." Doesn't this statement contradict the softened, "on-the-fence" statements earlier in the review paper? Is the formation of planetesimals the only conceivable route to planet formation? The review paper described here is a state-of-the-art summary of research on the formation of terrestrial planets. It does not mention the previous paper I described that reported the conflict of observation with theory, probably because the latter was published at about the same time. This is indeed a rapidly advancing field.

So, how do books aimed at the layperson handle the fact that current theories about planet formation are founded upon an impossibility? Whilst scientific papers are fair game for scientific scrutiny (they *should* be rigorously written to defend their arguments), books aimed at the layperson are not necessarily written in a scientifically defendable form. I don't want to drag individual authors into the mud, so you will have to judge for yourself. Take a look at various books on the subject, and read them with a critical and questioning posture. There are generally two approaches that authors of such books take. One approach is to not even mention the problem, in the hope that the reader doesn't realize that gravity cannot stick together objects that are not massive enough, especially in a turbulent environment. The author typically ends with the conclusion that the core accretion theory of planet formation is acceptable. In the other approach, the gist of the arguments given is that scientists are working with some wonderful computer codes that are making dust-sized particles stick together. The fact that rocks, boulders, and mountains coming at each other at tremendous speeds prevent any build-up of mass is marginalized (almost ignored) and the author typically moves straight on to gravitational accretion and growth of much larger masses (planetesimals). In other words, the critical failure of the entire scenario is skipped (or heavily downplayed), with the implication that people will progressively incorporate more and more complex physical processes into even more wonderful computer codes that will succeed in making the progression from dust to planetesimal. The account usually ends with something along the lines that there certainly are problems, but the general overall picture is correct. We are *nearly* there. Just a few loose ends, you understand. The account is inevitably tied together with generous handwaving. Underpinning all this is a thing called *hope*. The hope that somebody will find a way, so that the whole theory of planet formation won't have to be rewritten from scratch. The *hope* that something completely different, something that nobody has yet thought of, is *not* at the heart of planet formation. But hope is not a scientific procedure. Nor is belief. Scientists who think that the current theory of planet formation is essentially correct, *believe* that the "no-sticking" problem will be solved. It is a *belief* that is not based on scientific facts. It is true, if the problem were solved, the theory would agree with certain observational facts, but it would *still fail* to explain other observational facts (such as the planet desert mentioned earlier). But nobody knows whether it *can, even in principle*, be solved, so until then, supposing that it can, is a *belief*. Ultimately, data and observation will reveal the truth, so in some sense it doesn't matter. However, it does immensely slow down progress towards that goal.

The theory of planet migration (from larger to smaller distances from the host star) is intimately tied to theories of planet formation. Planet migration is an entirely theoretical expectation from the physics of the interaction of a planet with its environment. The processes involved are very complex and the study of planet migration constitutes a subfield in itself. Of course, a planet has never been observed to migrate. However, the finding of large numbers of hot Jupiters close to their host star, combined with the expectation that most planets are created near or beyond the snow line, is taken as evidence that the gas giants created there have migrated to the close-in positions. In the current paradigm, gas giants are created by accretion of hydrogen and helium onto a rocky core that has been created by means of the standard core-accretion scenario. The evidence for planet migration is therefore circumstantial and the conclusion (that some planets have migrated) relies on assumptions that may not be true. When planet formation is finally understood, it can be expected that an understanding of planet migration will also go hand-in-hand.

The top-down scenario of planet formation is known as the "disk instability model." In this scenario, instabilities in the protoplanetary disk cause large lumps of matter to break off and then contract under their own gravity to form planets. It is actually not straightforward to test this scenario because very complex physics is involved and correspondingly complex numerical computer codes are required to churn out the answers (with the associated large number of knobs and dials). The general conclusion appears to be that planet formation is possible under some circumstances, but only far out in the disk at tens of AU from the host star (recall that in our solar system Jupiter is nominally at about 5 AU from the Sun). There are other problems, such as the instabilities causing spiral arms to form in the disk instead of breaking up into clumps, and mass being transferred to the host star instead of going into planet formation. [13] The disk instability model is generally dismissed in the scientific literature. In a 2010 review paper on the topic, the authors concluded, "The ultimate fate and survivability of planets formed at such an early phase of disk evolution is unclear." [14] Then they state, "The principle of parsimonious explanations makes appeal to more than one formation mechanism for gas giants unappealing." This is extremely bizarre. This kind of statement has no place in science, but unfortunately it is not uncommon in peer-reviewed scientific literature. The authors are essentially saying that they *don't like* to have two modes of planet formation, based on a completely fictitious "principle" (if their "principle" were true, the Universe need not bother to exist at all since that would be very economical indeed). If the authors are using "parsimonious" as in "Ockham's razor" and therefore don't feel the need to further justify their statement, it is important to remember that neither the "law of parsimony" nor "Ockham's razor" is a "law" of nature or physics. In any particular case, nature and physics are not obligated by such inventions of the human mind. Nevertheless, it is still true that the disk instability model is not by itself a viable scenario for planet formation. [15]

Some researchers have actually begun to investigate an approach that is a hybrid of the bottom-up and top-down scenarios. An oversimplified explanation of this is that a gaseous "planetary embryo" forms by fragmentation far out in the protoplanetary disk (at 100 AU or so), and dust and other heavy elements contained in the embryo sink to the core (i.e., in a process of sedimentation). Thus, a rocky core is formed without running into the "sticking together" problem. The embryo migrates towards the host star, settling to become a gas giant, or if it migrates further, the gaseous envelope is stripped off (by differential gravitational, or tidal forces and evaporation), leaving a rocky terrestrial planet. The calculations are difficult and complex and the researchers working on these models themselves state that many details have yet to be addressed in order to see if it all works. [16] In particular, why should there be a pileup of hot Jupiters very close to the host star (at a distance that can be as small as 0.04 AU), as described earlier? In other words, if giant planets come in from 100 AU or so and lose their hydrogen and helium envelopes to become "naked" terrestrial planets closer to the star, how did the hot Jupiters make it to a distance that is even closer to the host star, *without* losing the envelope?

I'm going to make a suggestion that is just an idea, and not even a hypothesis. What if the very close-in hot Jupiters are actually planets being *born* (or recently formed), instead of planets headed to their death? What if planets are formed by a mass ejection from the host star? Such ejections would have to be more massive than those observed in the Sun in our epoch but observations do not rule it out. Jupiter has only 0.1% of the Sun's mass. A blob that has a few times this mass, would, after being ejected, try to hold itself together with its own self-gravity. Depending on its size and density, it would lose the outer layers until it was either disrupted completely, or until it attained a size and mass for which its self-gravity is able to hold on to the outermost layers of its gaseous envelope. This would account for the low density of the hot Jupiters that are close-in to their host stars: these exoplanets appear to be close to tidal disruption (as discussed earlier) not because they are being torn apart, but because they are *in the process of forming*. At the same time, heavy elements in the blob would be sinking to the center of the planet due to gravity (sedimentation), forming a rocky core. The overall, average density could still be small if the high-density core occupies a relatively small volume. Recall that in one of the most well-known hot Jupiters (HD 209458b) a tail-like structure has been claimed to be observed, trailing from the planet. As the ejected blob moves outward, it could lose all of its hydrogen and helium envelope, to become a terrestrial planet. It would settle at a distance determined by the initial energy of the impulsive ejection (as well as other factors). Very energetic ejections could conceivably send the planet out further, and it is possible that the terrestrial planet would again start to accrete hydrogen and helium as it plunges through ices and other material in the protoplanetary disk, eventually to settle as a gas giant or an ice giant.

A paper published in *Nature* argued that the close-in hot Jupiters could not be formed there, near the host star, but must have migrated in from a much larger distance. [17] Two arguments are given in that paper, both of which can be dismissed after the second paragraph in the paper. The first argument is that planets form in the protoplanetary disk and the disk is predicted to be too hot at the location of the hot Jupiters for small solid particles to start sticking together. Well, it is not known that the planets we observe were formed from a disk in the manner that is supposed, and, as explained earlier, we know that the scenario of planets being built up from small particles does not work. So the first argument in the paper can be dismissed, as it is not based on facts or observational evidence. The second argument given in the paper is that evaporation of the planet would have been a problem in the past, although it appears to be small now. However, this argument does not actually provide any reason to dismiss planet formation close to the host star. There is no reason why we could not start off with a larger mass, and end up with a hot Jupiter after mass loss.

So neither of the two arguments provides any reason why planets could not be formed from mass ejections from the host star. Nevertheless, there are likely to be plenty of other reasons why the ejection scenario would not work. However, some variation or modification of it might work, and it merits further investigation, considering that all of the current planet-formation scenarios are *known* not to work all the way (from start to finish).

As I have already mentioned, the close-in hot Jupiters present a problem for the current paradigms of planet formation and migration *especially* from the point-of-view of misalignment between the orbital plane and the host star's spin, not to mention the occurrence of highly eccentric orbits. [18] It is interesting that a 2010 study of the spin-orbit problem in hot Jupiters concluded that, "At present, standard disc (*sic*) migration cannot explain the observations without invoking at least another additional process." [19] It is also interesting that while some studies of the tidal stability of close-in hot Jupiters conclude that these exoplanets are falling in to the host star to eventually be destroyed, it has also been demonstrated that it is the youngest host stars that entertain the closest-in hot Jupiters. Given that the "falling" is *not* deduced from a measured direction of the motion, but simply from the fact that the star's gravity overpowers the planet's surface gravity, an alternative view is that young stars are more likely to undergo significant mass ejections, and that the hot Jupiters are freshly created and *are on their way out*, and are not falling in. [20]

A host-star ejection scenario for the creation of hot Jupiters simply does not encounter the spin-orbit misalignment problem or high eccentricity problem. The ejection does not have to occur in the plane of the protoplanetary disk so planets are not constrained to be created in that plane in the first place. On the other hand, we would have to explain why many planets *are* found to be close to

132

the plane of the disk.

How Common are Planets around Stars?

This is actually not an easy question to answer. The question cannot simply be answered by dividing the number of exoplanets found, by the total number of stars that are surveyed. All of the possible observational biases have to be studied in detail and then corrections have to be estimated for the exoplanets that might have been missed due to observational bias. Often, the observational biases must be studied in order to design the survey strategy itself. For example, velocity measurements of stars are more difficult for massive stars than stars with masses similar to the Sun. This is because the stellar rotation is faster for more massive stars, resulting in distortion of the Doppler measurements. There are also fewer measurable features for velocimetry of more massive stars because they are hotter and many of the elements' atoms are stripped of their electrons (whose transitions between atomic energy levels give rise to the measurable signatures). Several studies designed to establish the planet incidence rate have been conducted, and the dependence of the incidence of planets on other things such as the relative proportion of heavy elements, or stellar mass, has also been investigated. [21] The studies have been conducted over a range in stellar masses, from about a third of a solar mass to three solar masses. It is found that planet incidence increases from about 2% for the lowest stellar masses, to about 9% for stellar masses greater than about 1.3 times the solar mass. The numbers do depend on the detailed assumptions, however. Understanding the planet incidence rate is of course tied to an understanding of planet formation, which as I have already discussed earlier, lacks a satisfactory theoretical understanding. In addition, the planet incidence rate is sensitive to other assumptions as well, such as the dependence of the position of the snow line on the stellar mass. (Although the definition of the snow line as the boundary between liquid and solid water phases seems like it should be straightforward to calculate, it is not. A complex physical situation must be modeled and various assumptions need to be made.) [22]

While on the subject of the dependence of exoplanet incidence as a function of stellar mass, it is worth pointing out an intriguing and tentative result. That is, gas giants (e.g., Jupiter-mass exoplanets) appear to be far less likely to be found to be associated with low-mass cool stars than with high-mass hot stars that have evolved off of the main sequence (which is the main hydrogen-burning phase). In technical language, the stellar classification scheme that labels stars with the sequential letters O, B, A, F, G, K, M, very roughly indicates the temperature of a star (with "O" being the hottest, and M being the coolest). It is really a distribution of energy with wavelength (the spectrum) that is indicated by the code, but this has a rough correspondence with surface temperature. The letter code is thus also known as the "spectral type" of the star. Since the stellar luminosity is a function of surface temperature and surface area, and luminosity is dependent upon the stellar mass (from the physics of stellar structure), the spectral type (letter code) of a star is also an indicator of mass. Thus, there are stars known as M dwarfs that are amongst the least massive and coolest main sequence stars. Our Sun has a spectral type of G. The massive stars that were at the hot end of the range during their main sequence phase become cooler when they evolve off of the main sequence, but also become bloated in size, and are referred to as giants or subgiants (obviously depending on their size). An advantage of studying such evolved stars is that when they were too hot and on the main sequence, velocity measurements due to planets would have been difficult, but when they have cooled down some elements regain the electrons that were previously stripped off, providing more observables to make higher-precision velocity measurements. Returning to the question of the occurrence of Jupiter-mass exoplanets, a study in 2008 found that gas giants are far more likely to be found to be associated with more massive stars, in long period orbits, rather than with M dwarfs. Specifically, the study found that, compared to F, G, and K stars, M dwarfs are 3 to 10 times less likely to harbor a gas giant with an orbital period of less than 2,000 days. These conclusions were made after correcting for known observational biases. [23] Furthermore, it appears that stellar giants and subgiants are even more likely than F, G, or K stars to harbor long-period gas giants (although exceptions definitely exist and need to be explained). This trend is in the opposite sense to that which would be expected from an obvious observational bias that would favor the detection of short-period planets over long-period ones (because the short-period planets would be closer to the host star and produce more "wobble"). Finally, another observational fact is that no exoplanet with a mass comparable to Jupiter has been found closer than 0.1 AU to an M dwarf. [24] Studies that try to improve our understanding of the incidence of exoplanets as a function of the stellar parameters and evolutionary state are still ongoing.

Correlation of Planet Incidence with Heavy Element Abundance

Another intriguing finding that has become apparent in recent years is that exoplanets seem more likely to be found to be associated with metal-rich stars than metal-poor stars. By metal-rich and metal-poor I refer to the relative proportion of heavy elements (atomic number corresponding to carbon or higher) compared to

135

hydrogen and helium. Independent studies have attempted to eliminate selection effects in different ways and they come up with the same conclusion. [25] There is still some debate as to whether all observational biases have been eliminated. In particular, the metal-poor systems tend to be located in a part of the Galaxy that is known to be rich in hydrogen and helium.

The interpretation of the result that the incidence of exoplanets is higher around metal-rich stars is ambiguous. Is it because the star has been enriched in heavy elements by giant hot Jupiters migrating and spiraling into the star? Or is it because planet formation is more efficient if there is a higher concentration of heavy elements (naturally, more solid bodies can then be made)? The answer is unclear and the metallicity versus planet incidence correlation is a subject of active research.

A study in 2011 even proposed that a deficiency of refractory elements compared to volatiles in the host star is a signature of terrestrial planet formation. Here, the dividing line between refractory elements and volatiles is a condensation temperature of approximately 600 degrees Celsius, or 1080 degrees Fahrenheit (actually 900 degrees Kelvin). Our Sun is deficient in refractory elements compared to about 85% of most nearby "solar twins." The study speculates that the deficiency in our Sun compared to these solar twins came about because of the formation of terrestrial planets (wherein the missing elements lie). However, it has been found that solar twins that host hot Jupiters do not show the underabundance in refractory elements, and a majority (70%) of stars that have a composition more like our Sun do not have a close-in gas giant. This means that the proposed scenario of refractory element deficiency as an indicator of terrestrial planet formation implies that the presence of a close-in hot Jupiter must somehow prevent the formation of terrestrial planets. Clearly, more data is required to test predictions, and the authors of the study acknowledge that the proposed scenario is only a hypothesis. [26]

Multiplanet Systems

In the October 2011 exoplanet sample, the number of multiplanet systems is 81 (check the *Extrasolar Planets Encyclopedia* for the latest number). It is not known what fraction of stars contain multiplanet systems as opposed to just one planet because other planets may have gone undetected in single-planet systems.

It appears that many of the super-Earths and Neptunes that have been found reside in multiplanet systems. However, this could be an observational selection effect because low-mass planets require more observing time to build up statistical accuracy (since the signals are weaker), but multiplanet systems get more observing time because they get more attention from observers and selection committees because people think that multiplanet systems are more interesting than the single-planet systems.

Multiplanet systems have the particular characteristic that the orbits of the different planets in the system can interact with each other (by gravitational influence), and thereby become synchronized into resonant orbits. Several multiplanet systems have been found in which pairs of planets have orbital periods that have a ratio that is a whole number, or close to a whole number. [27] When such resonant orbits are found, the number of parameters that describes the system is reduced and the problem of solving for the orbital parameters can be overspecified. It is then possible to derive an inclination angle for the plane of the orbit and absolute planet masses (not just lower limits).

A Summary of the Important Findings So Far

I have described some of the essential new findings that have resulted from the exciting new science of exoplanet research. Often, these findings are in the form of new questions that we are faced with. Some of these questions were not even conceived before the flowering of the field only twenty years ago, but this in itself makes the endeavor all the more exciting. Although more tentative findings and correlations have been reported in the literature than I have discussed, I have tried to restrict the scope to those findings that are robust enough to survive the test of time. Tentative correlations that have a low statistical significance may not turn out to be real.

The discovery of so many "hot Jupiters" demands answers to fundamental questions. Of the 692 exoplanets in the October 2011 sample, 126 (about 18%) have a mass greater than half of that of Jupiter, and an orbital size (semimajor axis of the ellipse) of less than a tenth of the Earth-Sun distance. The corresponding orbital periods lie in the range of about 0.09 to 11 Earth days. If we consider the subsample of 192 exoplanets that have radii measurements, 109 of them satisfy the above mass and size criteria, and that constitutes about 57% of the subsample. Many of the close-in, hot gas giants have a density that is too low for simple models to account for, and this is a profound mystery. Why these underdense gas giants should be positioned to be at the verge of breakup by the tidal effects of the host star's gravity is also a mystery.

On a related note, there seems to be a distinct paucity of planets that have an orbital period that is less than 3 Earth days, whilst simultaneously having a mass in the range of about 1 Jupiter down to about 3% of Jupiter (or about 318 down to about 10 Earth masses respectively). This cannot be satisfactorily explained by observational bias because planets with a mass *less* than 10 Earth masses *and* a period of less than 3 days *are* found.

Giant planets are also found further away from the host star than the hot Jupiters, and ice giants (massive planets beyond the snow line) are common as well. The position of the snow line is not easy to calculate in the face of missing information, but we can say that of the full October 2011 sample of 692 exoplanets, 250 exoplanets (about 36%) have a mass greater than half of Jupiter and are located further than 1 AU from their host star. In all, nearly 70% of the exoplanets in the full sample have a mass greater than half of that of Jupiter, going up to about 31 Jupiter masses. However, it should be borne in mind that some of these proportions may be skewed by observational biases. Regardless of selection effects, theories of planet formation and planet migration must explain the properties of all the exoplanets found, especially those of the hot Jupiters. Moreover, selection effects cannot account for the abundance of hot Jupiters.

The bimodal distribution in the orbital period of exoplanets is an interesting problem that awaits a satisfactory explanation. Why is there a "planet desert" in the range of orbital periods of about 10 to 100 Earth days?

Early findings from studies of the fraction of stars that harbor planets, and attempts to determine which properties of stars favor hosting planets, have revealed a correlation of planet incidence with the relative proportion of heavy elements in the host star. The element abundances cannot yet be measured for the planets directly, they can only measured for the host star. Stars with a higher concentration of elements "heavier" than hydrogen and helium indicate enrichment either from "consuming" planets, or from the fact that some of the material used to make the star was already made in an earlier generation star and recycled. It is not known which of these is the correct interpretation. In addition, tentative results indicate that the type of star (its mass, temperature, size, etc.) makes a difference as to whether it is likely to host a planet or planets.

The proportion of exoplanets that have "squished" orbits compared to a circle (i.e., highly eccentric orbits) is much larger than was expected, and this, along with some of the other findings, does not fit in neatly with current ideas about planet formation and planet migration. Planets with orbits that are skewed out of the plane, compared to the plane of rotation of the host star, have been found to be common. Moreover, planets that orbit the star the "wrong" way around compared to the star's rotation are not insignificant in number. All of these things must eventually be explained naturally by correct theories of planet formation and migration.

Cold and hot Neptunes have been found, as well as super-Earths, but not in as great quantities as the giant planets. In the full October 2011 sample, 74 out of 686 exoplanets with mass estimates have a mass estimate in the range 10 to 50 Earth masses inclusive (our Neptune has a mass of about 17 Earth masses). That corresponds to about 11% of the October 2011 sample. A further 56 exoplanets have a mass that is less than 10 Earth masses (about 8% of the sample).

The October 2011 sample has only 4 exoplanets with a mass lower limit that is less than 2.5 Earth masses, but they are all very hot (closer than a third of the Earth-Sun distance to their host star). Earthlike planets in the habitable zone have yet to be found (but see discussion on the habitable zone).

We looked at a key study that captures the need for a paradigm shift in the planet formation process and in ideas about planet migration because the current scenario predicts a paucity of exoplanets in the regime of 5 to 30 Earth masses and orbital periods less than 50 days. Instead, observations show an overdensity of exoplanets in that regime.

Finally, it is worth pointing out what has *not* yet been possible. No moons have been found around exoplanets, mainly because of the sensitivity required. One method, the so-called gravitational lensing method, could in principle find them, but this observational technique is highly expensive in terms of the amount of observational time and resources required, and only 13 exoplanets were found by the beginning of October 2011 using the method (as described in chapter 3). One benefit of finding exomoons would be that of obtaining a better constraint on the mass of the planet that the exomoon is orbiting (from measurements of the orbital parameters of the exomoon).

Tentative Results from the Kepler Candidates Sample

Although I have said that I am going to stick with the findings that are robust and not too tentative, I will briefly summarize what has been found from the sample of 1,235 exoplanet *candidates* detected by the *Kepler* mission because it's worth taking a sneak peak. At the beginning of October 2011 there were 24 confirmed *Kepler* exoplanets, and these are in addition to the 1,235 candidates (check the *Kepler* website for the latest numbers). The early results have been succinctly summarized in a paper by the *Kepler* team. [28] The paper warns that although many false detections have been rejected, after more follow-up work, some false positives may remain (i.e., some of the exoplanet candidates will not turn out to be exoplanets). The paper also warns that the various observational biases have not yet been fully quantified.

The candidate sample is then cut by not including exoplanet candidates that have a size (radius) larger than twice that of Jupiter.

This is because very large exoplanet candidates are at a greater risk of not being exoplanets because they could turn out to be brown dwarfs (large blobs of matter that could have been stars but weren't massive enough to ignite nuclear burning of hydrogen). The size cutoff corresponds to roughly eight times the mass of Jupiter (for an exoplanet with the same density as Jupiter). The reduced sample then contains 1,202 candidates.

Remember that the transit method of exoplanet detection that *Kepler* employs cannot measure the mass of an exoplanet (or even a lower limit). Therefore only the sizes of the exoplanets are known and nothing is known of their masses. To get masses requires follow-up work. The paper goes on to classify the various kinds of categories of exoplanets, but this can only be done by a size comparison to the planets in our solar system. By far the largest size range found is the 2 to 6 Earth-size range, and these are referred to as Neptune-size exoplanets. These constitute about 55% of the sample. Next, the Jupiter-sized exoplanet candidates, from 6 to 15 Earth radii, constitute just under 14% of the sample. There are 19 out of 1,202 candidates (about 1.6%) that are very large, in the range 15 to 22 Earth radii. Going to candidates smaller than Neptune size, about 24% of exoplanets in the sample are super-Earths, having a radius of 1.25 to 2 Earth radii. If these candidates had the same density as Earth, they would have a mass range that is roughly 2 to 8 times the mass of Earth. Finally, 68 of the 1,202 candidates (about 5.7%) are smaller than 1.25 times the size of Earth. Overall, nearly three quarters of the candidates (74%) are smaller than Neptune. The size distribution has a maximum at around 2 to 3 Earth radii, which is interesting. It should be noted, however, that the above proportions are *not* corrected for observational biases (i.e., the sensitivity to different sizes of exoplanets is not the same). Once these sensitivity corrections are included, the occurrence frequencies are reported (as a percentage of the number of star systems examined), to be: 6% for the Earth-size category, 7% for super-Earths, 17% for Neptune size, and 4% for Jupiter size. A tiny percentage (about 0.02%) is estimated in the category containing candidates larger than twice the size of Jupiter.

This is all quite astonishing, because it means that exoplanets could be found in up to 34% of all star systems. How much this number goes down, after follow-up work on attempting to confirm each candidate, is uncertain. It is also worth pointing out that 17% of the host stars harbor more than one exoplanet candidate, and about 34% of all candidates are part of a multiplanet system.

Only 54 of the 1, 202 candidates are identified as being *potentially* in the habitable zone. Calculating whether an exoplanet is in the habitable zone is complex and carries many caveats. As explained earlier, one of the major uncertainties is the temperature, and the temperatures calculated for these candidates should only be interpreted as nominal, as it depends on many assumptions, which may not hold. I will say more about the habitable zone later, but for the moment I point out that it appears that the size of the exoplanet candidates in the potentially habitable zone ranges from a size smaller than Earth to a size greater than Jupiter. Of the 54 candidates, only 5 are actually smaller than Earth. The exact definition of the habitable zone that is adopted, along with the related assumptions that go into estimating temperatures, will obviously change these numbers.

Sorry, that was a high intensity of numerical information. However, it's good to get a feel for the demographics and properties of the exoplanet population that is revealing itself to the human race, after being hidden for so long. In a strange but very real way, the Universe is discovering *itself*. Revealing itself to itself. What an absolutely outrageous concept!

Before we leave the *Kepler* exoplanet candidate population, one more thing that stands out is that there is a prominent decrease in the number of exoplanet candidates, regardless of the size of the candidate, in the regime of an orbital period of less than 3 days, and a semimajor axis of less than 4% of the Earth-Sun distance. This same effect is, as I have discussed earlier, found in the exoplanets that were not discovered by *Kepler*.

Finally, it is imperative to bear in mind that the sizes of the exoplanet candidates may be wrong by a large amount (as much as a factor of 2, or 100%). One reason is that the host star size (on which an exoplanet size estimate depends), is not well-known before follow-up work is done.

Chapter 5

Where Is Everyone?

What Is Life?

This book is principally about exoplanets, and a discussion of the prospects of finding life on other planets and what we might expect would occupy another book, at least. On the other hand, for most people, including myself, the ultimate goal of studying exoplanets in the first place is to position ourselves so that we can look for, and hopefully find, life on other planets. Let's face it, studying exoplanets just for the sake of learning about rocks and gases orbiting stars would be pretty boring if we stopped there. If we extrapolate to the most favorable scenario in which humans actually make contact (eventually) with intelligent life-forms outside of our solar system, there is a vision that surely every person on this planet would show at least some curiosity. Making such a contact with another intelligent portion of the Universe would be a remarkable culmination of the collective stream of consciousness of the human race that has grown for millennia. An activity that presumably started with exploring our immediate little neighborhoods and trying to come to grips with the meaning of our self-awareness.

So I will briefly venture into this formidable territory, in the restricted context of bringing you up to speed with where we stand, in particular in the light of the recent rapid advances in the study of exoplanets and other relevant disciplines.

As soon as we turn our attention to life on other celestial bodies, we immediately run into the first problem. That is, "what *is* life?" You are probably aware that this question has vexed anyone who has asked it for a long time (obviously we cannot say by whom or when this question was first conceived). The problem is that it is always possible to "pick holes" in any proposed definition of life, because the essence of what life is still eludes us. I will not get into a discussion of the various definitions that have been proposed because it is not useful. There is no real consensus. Furthermore, I don't think we will be able to define life until we understand exactly how life originates from the beginning (starting with simple molecules). We will therefore put aside the question of what exactly life is and see how far we can go.

How does it make sense to talk about something when we can't even define what it is that we're talking about? Well, we do know various things that any theory or proposition must, at a minimum, explain about life-forms on Earth that nobody would contest are "alive." In other words, we must at least be able to understand the origin of some known life-forms on Earth. In any case, the best we can hope to do at the moment is to get a basic idea of what the key issues are since so much is unknown and not understood.

The Origin of Life

Discussions on the origin of life invariably center around problems associated with making molecular building blocks of life such as amino acids, proteins, enzymes, going on ultimately to making RNA (ribonucleic acid) and DNA (deoxyribonucleic acid) out of simple molecules that are found in nonliving matter. [1] The circularity in the arguments enters right here because we can ask, "What exactly is a simple molecule?" Is an amino acid simple? By itself, an amino acid is not a living thing. However, the necessary existence of the 20 amino acids that are the building blocks of more complex molecules that we know are characteristic of living things, is a prerequisite for life as we know it on Earth. However, what some people may not realize is that the basic problem of how the "code" to make a living organism out of simple building blocks, without instructions or a blueprint to begin with, is a fundamental problem that is *independent of specifics of the chemistry and biochemistry*. Too often, discussions on the origin of life get bogged down with the complications associated with the specific details, when really the essence of the problem is much more general. Currently, *all* attempts at explaining the origin of life try to generate some kind of order out of chaos, relying on rare random events (whose probabilities are usually impossible to estimate) that somehow get favorably selected and replicated. This is the first hurdle, because it is not the natural order of things. That is, the *entropy*, or amount of disorder of a closed system, always increases (an observational fact embodied in the "Second Law of Thermodynamics"). It is possible to force things to go "in the wrong direction" in a local part of the Universe, but to do that requires an input of energy, a device or mechanism to do it, and an overall increase in entropy elsewhere in Universe.

An example of a device that *decreases* the entropy in a local part of the Universe is an air conditioner. This device makes heat flow in the wrong direction. An equivalent statement of the law of increase of entropy is that heat flows from hot things to cold things and not the other way around. Heat is a kind of "last stop" for the degradation of information because it is a state of maximal disorder. Therefore, an air conditioner is something that takes in energy as an input and works to force heat to flow in the wrong direction, thereby making entropy decrease instead of increase, in a restricted region of space. That region becomes more ordered and cooler than the surroundings. If you have stood by an air conditioner you know that it pumps out heat to the surroundings. Overall, taking the cooled region and the surroundings together, we find that the overall entropy of the Universe still increases (and therefore the Second Law of Thermodynamics is upheld). To be sure, there are many physical situations in which a local decrease in entropy (or an increase in order) can happen spontaneously, and the freezing of water into ice is a common example. However, such scenarios don't come anywhere close to meeting the requirements of an origin of life.

The point of the discussion about air conditioners is that if you demand that a living entity be created from prebiotic molecules by means of rare *statistical fluctuations* amongst random events, it is equivalent to demanding that an air conditioner is created from random events involving its component parts, *and* that this air conditioner is able to make copies of itself, *and* that this air conditioner is able to utilize materials in its surroundings for its power supply. The air conditioner is to be created without any plan or blueprint, because only random events combined with selection are allowed in current "theories" of the origin of life. However, *before there is any replication capability, there can be no selection process*. This is only one of several fundamental problems that any explanation for the origin of life faces, without having to invoke new, as yet unknown, physical principles. An entropy-reducing device is supposed to be built out of disorder "by itself." The local entropy-reducing aspect is a common factor of all things that are undisputed as living entities. If this seems unlikely, you are right, it is.

I will save you some trouble and tell you that nobody has demonstrated that this is actually how life began on Earth from prebiotic matter. Nobody has actually demonstrated that it is even possible, given the stupendously small probabilities, and the finite age of the Earth (billions of years is a long time but it is very short when stacked against virtually zero probabilities). However, you will get a different story depending on which book you read, or whom you listen to. At one extreme you will be told that it is as good as "fact" that life originated as a result of statistically unlikely events, and at the other extreme you will be told that we actually have no clue about the origin of life. Both conclusions are based on the same mathematical and physical evidence. I will give you more details (and citations) of these diametrically opposing views later when I discuss the specifics of life on Earth.

Some scientists believe that life is simply a consequence of increasing the complexity of a system beyond some critical threshold. [2] School textbooks particularly like to portray this nebulous view and stop there. The notion is attractive but has no factual foundation. The space shuttle is an extremely complex system, way more complex than some simple life-forms, but the space shuttle is not alive, so the viewpoint is not even true. Clearly there is something in addition to complexity that living entities have. Texts that invoke the notion of life originating as a result of complexity never propose any mechanism that causes the origin of life. Statements to the effect that complexity alone gives rise to life are not helpful and are in fact excellent *nonexplanations*.

It should also be clear that something in addition to the creation of order out of disorder is required. Presumably, an ordered system has to *do* something in order to be considered to be alive (as opposed to, for example, a crystal, which is a highly ordered state of matter but doesn't really *do* anything).

There does, however, seem to be some general agreement on two things that are necessary to incorporate into any explanation of the origin of life. These are *metabolism* and *replication*. Here we again run into problems because when we begin to think about defining what metabolism is exactly, it turns out that it is currently impossible to define it in any precise way, and there is debate about how it should be defined. Freeman Dyson, an expert who has spent considerable effort trying to understand the origin of life, refers to this bizarre situation quite frankly in his book, *Origins of Life*. Just before he launches into a mathematical exploration of a very restricted hypothetical scenario to investigate some properties of a system that makes a transition from disorder to a less disordered state, he explicitly states that metabolism will be defined *after the fact*. In other words, leaving open the types of interaction that are possible between subunits of a system, in this view, metabolism is "defined" as everything that the system ends up *doing* that is not labeled as replication. Dyson admits that in such a set up it is not possible to investigate the origin of metabolism. That is precisely because in this view metabolism is simply a description of what a system ends up *doing*, and it is not really a definition. To be sure, we know that at the very least, metabolism involves some kind of continuous pumping of entropy in the wrong direction, in a localized region of space. However, would you say that an air conditioner metabolizes? In the end it will probably be necessary to classify what different types of systems invented to model the origin of life ending up "doing" and then relate those classifications to what we actually observe in real living systems.

It should also be clear that in order for replication and selection to work (in terms of increasing the information content and sophistication of an entity that will give rise to a living organism), ordered states must be transient. In other words, death is a necessity. If living entities, or precursors to living entities, did not die, you would just end up with a big pile of living and nonliving goo consisting of things that "don't work." This latter term is of course subjective because humans decide what "works" and what doesn't because only humans invent an end goal. As far as nature is concerned, it doesn't matter. The point is that without death, *everything* just sticks around and there is no preference or advantage of one thing existing in favor over another. You will also notice that I have avoided defining what "random" and "information" mean exactly (it is not obvious). It is beyond the scope of this book to go into the physics and mathematics of that but for the purpose here, it is sufficient to say that if we are talking about information that describes all of the physical attributes and properties of a system, the higher the information content, the smaller the number of bits of data required to do the job.

Going back to metabolism and replication, you might think that it is obvious that replication must come before metabolism. In other words, to explain the origin of life, does a system need to develop a method of self-replication first, or does it need to metabolize first (whatever that means exactly)? There are problems with both scenarios, and the two extreme perspectives are known, in the context of the specific biochemistry of life on Earth, as the *Oparin hypothesis* (metabolism first) and the *Eigen theory* (replication first). The Oparin hypothesis dates back to A. Oparin in 1924, and the Eigen theory came later, in 1981. [3] According to F. Dyson, the Eigen scenario is "now the most fashionable and generally accepted theory." Of the Oparin theory, Dyson says, "it is unfashionable." [4] However, I note that I have come across school-level books that don't even mention Eigen. [5] The two scenarios mentioned here are not the only possibilities, but other scenarios are in the end some variation of the same basic principles, and no

151

scenario is satisfactory.

There are several very serious problems with a scenario that invokes replication first. One is the intolerance to errors in the replication. Before getting into the details, it is important to understand how prebiotic matter is supposed to start self-replicating in the first place. All current models are based on variation, selection, and evolution, the variations originating in chance events from statistical fluctuations. The replication is *required* to be imperfect in order for self-replication to develop in the first place. However, if the error rate is too high the errors will accumulate and eventually could become fatal enough to kill the entire population of replicators. Remember that the replicators cannot be immortal; death has to be present for them to exist in the first place. In fact, the Eigen scenario *assumes* that the variations remain bounded and that the selection process selects for objects that are close to the population norm or average in some sense. This is of course rather vague but it doesn't matter because this fine-tuning of the foundation of the scenario amounts to fudging by hand. Even with this cheating, the system cannot avoid catastrophe if the error rate is too large. The very thing that is the origin of replication is the same thing that can kill it.

Three other problems in the Eigen scenario have been shown to be showstoppers by means of sophisticated computer simulations. [6] It is not clear to me why it took computer simulations to reveal these problems because they are transparent to a nonexpert applying some basic common sense. In the Eigen scenario, the replicating molecule (RNA) serves several roles in the multiple steps that are required for a complete cycle of replication. One of those roles is that RNA serves as a catalyst, performing essential facilitation functions. However, it is possible that a single RNA unit evolves, due to chance events, to be able to repli-

152

cate faster than the other RNA units and stops acting as a catalyst. This new super-replicating RNA that doesn't bother catalyzing, rapidly swamps the soup and throttles the rest of the population to death. Why would you need a fancy computer simulation to realize this? It's pretty obvious. Another possibility is that the RNA unit evolves to start skipping steps in the chain, catalyzing later steps and missing earlier ones. The chain cannot be completed and we end up with a population that is completely different (in other words the original population ceases to exist). A third problem is simply that a statistical fluctuation in a critical component in the chain renders the quantity of that component to be zero. There is always a finite probability that this can happen and it results in a complete collapse of the population. Again, all of this can be deduced with some application of common sense rather than fancy computer code.

In the Oparin scenario, replication comes after metabolism but some of the issues discussed above for the Eigen theory are still problematic. The key difference is that the first stages of replication are very crude and in fact the kind of replication envisaged is not replication in the true sense of the word. The idea behind the Oparin scenario is quite vague and is based on the premise that there existed droplets of oily liquid containing a jumble of molecules. These "bags" of molecules are supposed to resemble, in some primitive and crude way, future living cells. It is then postulated that some chance events led to more highly organized states. Successive "jumps" are supposed to lead to progressively more organization in the bag of molecules. In the context of the particular biochemistry on Earth, this means that some molecules will, by pure chance, act as catalysts or enzymes that facilitate chemical reactions. Then it is further supposed that more chance events may lead to a splitting of the bag of molecules into two similar (but not necessarily identical) bags. This is supposed to be a precursor to cell division and more exact replication in the future. The implication is that the more organized states have to be more stable in order for there to be any preferential selection (otherwise all states would simply accumulate into a goo). One of the many problems with this scenario is that there is no precedent that successive generations have any continuity, since the replication is only crude. The only driver for the replication to become more exact is that one bag out of a huge number must be much more stable than its competitors, and the competitors must all die quickly and efficiently (why should they?). On top of that, there is no guarantee that a jump to a more ordered state will not be detrimental. Although there is a higher tolerance to a chance event bringing down the whole population (because replication is not exact), that is only true at the beginning. You are trying to create a system that is heading towards exact replication so by definition you will eventually run into the same problems that are associated with trying to develop replication from chance events alone. You do not need to be a Nobel Prize winner to figure all this out.

Another major problem with the Oparin scenario is that a more organized jumble of molecules does not necessarily mean that the bag is any closer to what you want to call a living thing. You could have a highly organized bag of molecules, but it might still not be living. Complexity in itself is not enough. The Oparin scenario makes no attempt to address this, and metabolism is simply a label for whatever the bags end up "doing." The Oparin scenario essentially amounts to unquantified handwaving. The Eigen scenario also requires events to have taken place for which there is no actual evidence and involves putting things in by hand. Yet these are the leading "explanations" for the origin of life. Is that really the best that can be done by the finest experts working on the problem? If you read about it in other books you will see a wide range in the way things are presented. At one extreme it is depicted with a great degree of certainty. For example, in the book *Schaums's Outline of Biology* (which adopts the Oparin scenario), [7] the best that is offered in the chapter called "Origin of Life" is as follows: "The organic material of the seas, becoming increasingly concentrated, accreted into larger molecules of spatial, or structural complexity- colloids with special properties of electric charge, adsorptive powers, translational movements, and even the ability to divide after reaching a certain size."

This has got to be the best nonexplanation of the origin of life that I have ever seen! It is a jumble of words strung together into a sentence that doesn't even begin to address the problems of the origin of life, leaving an overall sentence that is riddled with ambiguity and lack of meaning. At best it is simply a list of vague events that some people think might have happened and are in some way relevant to the origin of life. However, the list (and the entire chapter) provides absolutely no explanation of anything. The entire chapter in that book avoids any mention of the large number of assumptions that have to be made (even if you were able to resolve the ambiguities), and there is no discussion of the issues that challenge the scenario. In particular there is no discussion of *statistical noise*. The book wants the reader to believe that the assumption that increasing complexity is all that is required to produce life, but this is manifestly false. At the other extreme, Freeman Dyson gives a more honest assessment in his book, *Origins of Life*, in which he points out that the Oparin scenario has not been tested by computer simulations. The actual probabilities in the Oparin and other scenarios are impossible to calculate at the moment because nothing is known about the actual circumstances of the origin of life (and even if we had that information, some estimation and approximation would still be necessary due to the complexity of the problem). Dyson goes on to say that, "Concerning the origin of life itself, the watershed between chemistry and biology, the transition between lifeless chemical activity and organized biological metabolism, there is no direct evidence at all." Then, referring to the Oparin, Eigen, and a third scenario that is in a way intermediate between the two, Dyson concludes, "At present there is no compelling reason to accept or to reject any of the three theories. Any of them, or none of them, could turn out to be right." [8]

As I have said, all of the current scenarios for the origin of life are based only on probabilities. Until someone can think of a way that a particular scenario can be falsifiable, the scenario does not qualify as a scientific hypothesis and does not enter the realm of science. Otherwise, anybody could say anything they liked and call it science. A scenario that is invented by a scientist does not in itself make that scenario a hypothesis that can be said to join the realm of science. All current scenarios for the origin of life are therefore not in the realm of science, they are beliefs (until the hypotheses can be formulated to be falsifiable). They are beliefs that could turn out to be completely wrong.

Let me put things in perspective for you. Would you agree that a dollar bill is a less complex structure than a living cell (say, for the sake of argument, that of a bacterium?), in terms of its atomic and chemical constitution? In all current scenarios of the origin of life, the cell came to exist because of a large number of chance events. In these scenarios there is absolutely no distinction between something evolving that looks like life as we know it and something that looks nothing like anything we have ever seen before. Being simpler than a cell, the probability of evolving a dollar bill that is identical to a real dollar bill is actually higher than the probability of evolving a living cell. If you believe any of the current scenarios concerning the origin of life, then you also believe that there is a finite probability of evolving a dollar bill that replicates itself, and the probability of this happening can be higher than it is for a replicating cell. You can imagine the replicating dollar bill evolving a cellulose packaging to protect itself (like a sandwich bag or something), a feat that would be much easier to achieve than all of the achievements of a real living cell. You then also have to believe that there is a finite chance of finding an exoplanet that is littered with billions of dollar bills wrapped in sandwich bags. You can't have it both ways. If you believe that living entities arose out of prebiotic material purely by statistical processes combined with selection then you have to also believe in the possibility of finding the money-laden exoplanets. Does that sound ridiculous? It is. If you don't think it is ridiculous, you might be able to convince yourself that, concerning life on Earth, it only had to happen once. However, the "it" here actually refers to a vast number of chance events only happening once. Still, you might convince yourself that all of these things happened. You are free to believe that, but your belief would be just that: belief and not science, until you can think of a way to falsify it or test it directly.

I have no alternative to offer to you. It is better to admit that we have no idea how life began on Earth than to believe in magical, nebulous and unfalsifiable scenarios. However, I will tell you that it is possible that we are looking for the answer in the wrong place entirely. Allow me to make an analogy. Suppose you lived at a time before electrons were discovered, before electricity and magnetism were understood. You looked up at the sky and wondered about the origin of lightning. In order to understand the true origin of lightning you would have to discover the electron and make a whole bunch of other discoveries about electricity and the behavior of electrons. But an electron is at least 18 orders of magnitude smaller than the size scale of lightning. In other words, an electron is a billion billion times smaller than the size of a lightning streak. You could study lightning all you like, you could make all the measurements on lightning that were available to you before the discovery of the electron until you were blue in the face. *You will not discover the electron by studying lightning.* The true origin of lightning will elude you until you discover the electron. The electron was not discovered until 1897. The electron was discovered by J. J. Thomson in specialized equipment in a laboratory. These experiments were not made to study lightning. Moreover, a vast number of advances in science had to be made (unknowingly) that predated the discovery of the electron. A lot more work then had to be done after the discovery of the electron in order to gain a better understanding of electricity and magnetism. Only then was it possible to understand the true origin of lightning. Let me spell out the conclusion: *You will not discover the true origin of lightning by studying lightning.* In a similar vein, if you picked up a lodestone and wanted to understand the origin of its magnetism you would again have to discover the electron. Again, you could study the lodestone until you were blue in the face but you will not discover the electron by studying the object itself, and the true origin of magnetism will elude you until you do discover the electron. My point is that it may be possible that we could study the biochemistry of life and its molecular and chemical properties as much as we like but the true origin of life might never be discovered if the answer lies elsewhere. Something else that might exist in a completely different physical regime compared to the one where we think the answer should be, may need to be discovered first. That different regime may exist on a different spatial scale, and/or it may involve as yet undiscovered physics.

You may wonder how any undiscovered physics could exist without having been noticed before. Wouldn't we have detected something in experiments already? Wouldn't it involve new forces, or at least wouldn't it contradict existing measurements? In other words, how can there be any more room for new physics? There are two things that you should appreciate. One is that the boundary between physics, chemistry, and biology, the regime in which the complex molecules of life interact, is poorly studied, both observationally and theoretically. The second thing is that the uncovering of new principles and concepts in physics, without upsetting known measurements and observations, has happened in science many times over in history. It happened with the discovery of quantum mechanics at the beginning of the twentieth century, in many different situations. Prior to quantum mechanics, a longstanding problem in atomic physics was that, according to prevailing theory at the time, electrons in atoms should not exist in stable configurations, but they should instead be losing energy and spiraling into the nucleus to oblivion. As usual, the puzzle was given a nice name that ended in a buzzword ("catastrophe" in this case), implying that one day a way would be found to resolve the puzzle by finding the missing piece in the existing theory. However, nobody anticipated that the "loose ends" would actually lead to an entirely new branch of physics, overturning basic conceptualizations of the nature of matter. [9] Every physicist you speak to will agree that it was a major revolution. Quantum mechanics was discovered in the process of resolving a different (and very important) "loose end" and the result was to turn physics completely on its head. It was a new beginning for physics, not the end. Quantum mechanics resolved the problem of the electrons in the atoms. The electrons could not spiral into the nucleus as a direct consequence of the new theory. The important thing to realize is that

159

no existing measurements or observations were upset (how could they be?). On the contrary, previous measurements *confirmed* the new interpretation. The new theory was actually able to "predict" existing measurements that could not be explained before. Moreover, the new theory required a new entity to be invented, called a "wavefunction." Such a thing cannot be directly observed or measured, only indirect effects of it can be observed and measured. But the wavefunction appears in the equations of the theory and has specific mathematical properties (that depend on the context). These wavefunctions obviously were around before humans discovered quantum mechanics, and anyone could have argued before that discovery (as indeed many people did), that there was no room for any new physics. But they would have been wrong. In the problem of the origin of life, existing scenarios (I cannot call them theories) don't make any quantitative predictions, so discovery of new physics (or at least physical interactions that were previously unknown) will not upset any predictions (because there aren't any). I am not calling this idea a hypothesis, because it is not yet in the realm of science. It is simply a suggestion to stop hanging on exclusively to ideas that can't even be demonstrated to work and can't be falsified.

Challenges for the Origin of Life on Earth

So far I have talked about the very general problem of generating life out of nonliving simple units, without reference to what those units are because that general problem is the same regardless of what the units are. It can be summarized by saying that, at the very least, the instructions to build a self-replicating entropy-reducing device, and a mechanism to read the instructions and build the device, have to appear out of a chaotic goo made out of units that don't self-replicate or reduce entropy, all through a long chain of chance events (and the entire process has to overcome statistical noise). Selection processes, or any equivalent of "survival of the fittest," cannot begin to work before a system is able to replicate. Any life-form on any planet or other celestial body will be faced with these same problems and the issues discussed earlier.

Now we turn to the specific biochemistry associated with life on Earth and the specific problems in this context. This is just going to be an oversimplified sketch. First of all, we need the 20 amino acids that are used in various combinations, as the building blocks that make up all the proteins, enzymes, catalysts required for life. Some of them are also the components of the code in DNA and RNA. The amino acids are not large molecules and their three-dimensional structure is important as far as their functions are concerned. They are composed of various combinations of atoms, hydrogen, carbon, oxygen, and nitrogen being the most common. The name of the game is to then try to first make amino acids from simpler molecules that might have been abundant on the primordial Earth, and then ultimately to make the large and complex self-replicating molecule RNA, which plays a multitude of different roles in living organisms as we know them. [10] One of the fundamental problems in the general scheme of things is that in order to make proteins we need enzymes (to catalyze biochemical processes), but in order to make enzymes we need proteins. When it was discovered recently that RNA itself can act as an enzyme, the concept of the "RNA world" came about, which alleviates the problem to some extent. [11] However, there is still the problem of how a self-replicating molecule like RNA came about in the first place. Laboratory experiments are able to produce RNA from smaller molecular units ("monomers") using an enzyme without a template, and they can also produce RNA without an enzyme but with a template. Nobody has produced RNA with no enzyme and no template.

You may be thinking that there were reports (in 2010) in the news about scientists having made the first artificial living cell from scratch. It was nowhere close to that. There was a lot of razzmatazz in the reporting. The process was simply a *transfer*, not a creation. Already functioning DNA was inserted into an already existing cell. The point of that experiment was that the DNA of a cell could be replaced, potentially giving humans the ability to take control of a cell in some desired manner. So nobody has created a living cell out of lifeless chemical components, even with nonrandom processes.

Now let's go back to the amino acids, which are the building blocks of proteins and enzymes, and the "letters" of the code in DNA. Although it has been well-known for some time that amino acids can be delivered from space via meteorites, we really need to demonstrate that amino acids could have been made on Earth using locally available material at the time life is supposed to have begun. (Dozens of other, albeit simpler, organic molecules have also been directly observed in so-called "molecular clouds" in space.) You have probably heard of the famous "Miller and Urey experiment," which attempted to emulate lightning striking a primordial Earth environment, to see if amino acids could be made in conditions resembling the primordial Earth. The success of the experiment in producing a goo with a high yield of the amino acids has been rather overplayed, but even the "hope" from that is now undermined because the conclusions from that experiment are way out of date. In order to understand why, we need to look at what we mean by the terms *reducing atmosphere* and *oxidizing atmosphere*.

In chemistry, the term *reducing reaction* generally refers to the fact that the subject of a chemical reaction ends up losing negative electric charge (to the reducing agent). An *oxidizing reaction* is one in which negative charge is gained by the subject of the reaction (at the expense of the oxidizing agent losing the negative charge). Oxidation does not necessarily involve oxygen but it often does. A reducing atmosphere is then one that contains predominantly reducing agents, and an oxidizing atmosphere is one that contains predominantly oxidizing agents. Examples of constituents in a reducing atmosphere include hydrogen, methane, ammonia, and carbon dioxide. An oxygen atmosphere is of course oxidizing, and gases such as nitrogen and neon are neutral (i.e., they are neither reducing nor oxidizing agents).

Now the point of all of this talk about reducing and oxidizing atmospheres is that the Miller and Urey experiment only works in a reducing atmosphere. When the experiment is done with an oxidizing or neutral atmosphere, it fails completely. Up until recently, it was believed that the primordial Earth *did* have a reducing atmosphere at the time life is supposed to have originated. That epoch, about 3.5 to 3.8 billion years ago, was believed to be *after* the period of late heavy bombardment (by asteroids and similar objects) in the solar system. However, recent findings have changed all this. Studies of the radioactive species of carbon (the isotope carbon 13) in rocks in Greenland dated to be 3.8 billion years old reveal evidence that the carbon in these rocks has been biologically processed. [12] Since the period of heavy bombardment is supposed have ended about 3.8 billion years ago the implication is that some form of life existed during the bombardment period, and indeed survived it. Now it is believed that if there ever was a reducing atmosphere on the primordial Earth, it must have vanished by the time of the heavy bombardment period because of the existence of certain sedimentary rocks that have been dated and shown to include carbonates and various forms of oxidized iron. These could not have been formed in a reducing atmosphere. This means that the atmosphere at that time must have been neutral because free oxygen only appeared after life began, before that it was taken up to make water. The other thing is that neon is the seventh most abundant element in the Universe and it is more abundant than nitrogen. Yet the Earth's atmosphere today has a neon abundance that is only 1 part in 60,000 relative to nitrogen. The inference is that the original neon in the Earth's atmosphere was lost to space much earlier, when the temperature was too high for gravity to retain it. If the neon was lost in this way, other "lighter" gases such as hydrogen, nitrogen, and any ammonia would also have been lost in the same way. (Today's atmospheric nitrogen is thought to have been produced by volcanic activity.) The conclusion is then that the Miller and Urey scenario of producing amino acids out of primordial goo could not have worked, so it is no longer believed to be a viable mechanism. [13]

Instead of the Miller and Urey scenario, the picture that is now emerging (which may or may not be correct) is that life actually originated in the oceans around the warmth of deep vents. Remarkable discoveries of life existing on the Earth today in the most unlikely places have driven a paradigm shift in what we think are the conditions that are required for life. Deep in the ocean, in utter darkness, the weirdest creatures have been found, life-forms that defy the old preconception that light is critical for life. [14] I highly recommend the book *The Deep: The Extraordinary Creatures of the Abyss*, by C. Nouvian, which vividly depicts with beautiful photography some of the strangest creatures on Earth. If you want a glimpse at "exotic alien life," it's right there in the book. Some of the life-forms have transparent bodies, some of them generate colored lights, some of them have no eyes, and they all have remarkable body plans. These new discoveries are prompting the question of whether the surface of the Earth was even relevant for the origin of life. Life has been found not only several kilometers down in the deep ocean, but life has also been found to exist today in deep strata of underground rock. [15] Life-forms that exist in a variety of extreme conditions have now been found and are, appropriately enough, known as extremophiles. In particular, life-forms that exist in very hot environments, 80 degrees centigrade (about 180 degrees Fahrenheit) and higher, have been found in hot springs. [16] At the other temperature extreme, life-forms that exist in subfreezing temperatures in Antarctic sea ice have also been found. [17] In 2011, nematode worms were discovered in mines more than a mile below ground, a place previously thought to prohibit the support of multicellular organisms. [18] There is no light, no energy from photosynthesis, the temperature is very high, and oxygen is scarce. Yet they live. All of these findings raise new possibilities of life-forms existing on solar system members such as Mars, Titan, Triton, Europa, and others.

Getting back to the problem of the origin of life, we have seen that even making amino acids is problematic. To add to the issues already discussed, there is another unsolved mystery. It turns out that the geometrical arrangement of the component atoms of an amino acid molecule in three-dimensional space is critical for an amino acid to function correctly. In particular, there are two arrangements of an amino-acid molecule that are mirror images of each other but otherwise their chemical properties are identical. However, it turns out that in a biological system, the behavior and functionality of the two forms of amino acid are different. The two forms are referred to as left-handed and right-handed molecules respectively, because they cannot be mapped onto each other by a rotation or any other spatial manipulation. Which one is labeled left and which one is labeled right, is simply a matter of convention.

The property of "handedness" is generally known as *chirality*. In nature the two mirror images of a molecule occur in equal proportions. However, it is a remarkable fact that the molecules in living organisms are always left-handed. The right-handed counterpart cannot function with the left-handed apparatus. In a test tube, or in an organism that has died, left-handed (or right-handed) molecules revert to a mixture containing equal proportions of both types. The reason for the chirality of life is a complete mystery. No satisfactory explanation has been advanced. [19] Any theory of the origin of life has to explain the chirality problem.

If we acknowledge all of the problems in making amino acids, we can still investigate the problems that occur once we have the amino acids. If the goal is to make the nucleic acid RNA, we have to first construct the subunits, the nucleotides, which this long chain molecule is composed of. Those nucleotides consist of three parts that must be brought together: a base (an amino acid), a sugar, and a group containing the element phosphorus. There are more than a hundred ways that the three groups can be stuck together but only one of them is correct and functional. Even if the correct one was made by chance, you would have to figure out a way to get rid of all the wrong ones, otherwise they would swamp the correct one. Even if you could do *that*, it turns out that the correctly formed nucleotides are unstable in solution and quickly hydrolyze back into their component parts. This means that these nucleotides could not have been hanging around in primordial waters waiting to begin life. Nobody has figured out a way to make the nucleotides rapidly enough so that they can find each other and form the helical chain of RNA (which is not just a random collection of them). Moreover, the synthesis of nucleotides is difficult in all environments, reducing, oxidizing, and neutral. We can conclude by saying that experiments to produce nucleotides from elementary components have not been successful. Nobody has made RNA from its building blocks. The fact that it is unlikely that all the building blocks can be made by random processes from prebiotic material compounds the mystery.

Chemistry for Life: What's the Deal with Arsenic?

Life on Earth predominantly makes use of the elements hydrogen, carbon, nitrogen, oxygen, sulfur, and phosphorus. All of these elements are low down in the periodic table and have some of the highest abundances in the Universe. It is thought that life, if it is found anywhere else, is necessarily going to be based around molecules that feature carbon in a big way (as is the case on Earth). Carbon is unique in that, compared to other elements, its atoms can form more bonds with other atoms of its own kind, enabling the construction of very long chains and very complex molecular structures. It has been acknowledged that silicon atoms are also very good at forming bonds with themselves, and although not in the same league as carbon, it is conceivable that silicon-based life is feasible. Other than silicon, life based around any other element has not been given any serious consideration.

The element phosphorus is a key component of the molecule known as ATP, which is responsible for providing the energy source to cells. Phosphorus also appears in critical roles in nucleic acids and various proteins. In 2010, a team of scientists led by Felisa Wolfe-Simon discovered a bacterium in Mono Lake (California) that sustained growth by using arsenic instead of phosphorus. [20] It appeared that biomolecules, including DNA, were incorporating arsenic instead of phosphorus in order to function. This remarkable discovery provoked fierce reactions, both of surprise and anger. Surprise because arsenic is toxic to "normal" life.

Anger because the discoverers were accused of making the claim of arsenic incorporation into DNA based on impure samples containing the DNA. However, as of October 2011, the result has not been disproved either.

We should bear in mind that scientists have fallen prey, on more than one occasion, to drawing conclusions based on a sample of one (i.e., the one collection of known life on Earth). Considering that it is so hard to even define what life is, let alone explain its origin, it seems rather premature to conclude that the arsenic for phosphorus substitution is surprising. Surely toxicity is a relative attribute. If you consider a single atom, toxicity means nothing. If a life-form has been constructed from elements that *you* label as toxic (to things that *you* know about), why should it be surprising that the element in question is not toxic to the life-form that it grew up with? The surprise/anger combination in the reaction to the arsenic result had disturbing overtones of an emotional content that was distinctly excessive for scientific discourse. The result is not *that* controversial. Even a middle-school student knows that various groups of elements share key electrochemical properties and substitution of one element in the group by another to serve a similar but not identical role is not surprising. The reason for the overreaction is that people (not just scientists) generally do not like having their world view or "understanding" upset, challenged, and changed. The experimental results should absolutely be verified with a greater rigor and methodology before the result can be said to be fact. But surely it can be done in a courteous manner.

Water in the Moon

Before you start thinking that I've made a mistake in the heading, no I haven't. I mean "in the Moon" and not "on the Moon." In 2008 there were reports of water found in lunar volcanic glasses, indicating water in the Moon's interior, when it was previously thought impossible that there should be any significant amounts of water there. [21] A later paper, in 2009, subsequently reported the finding of surface water using a completely different method, along with measurements of hydroxyl (a combination involving one oxygen and one hydrogen atom). It was estimated that about ten thousand billion tons of water were deposited on the Moon by comets, over a period of two billion years, but that some of the water might be due to the delivery of hydrogen nuclei (protons) by the solar wind, combining with oxygen in materials on the Moon's surface. [22] The water and hydroxyl were detected all over the Moon's surface (in varying amounts) and it was estimated that the average abundance over the surface could be as much as 50%. The studies of water on the Moon's surface also hinted that there was some evidence to suggest that more water could be buried *beneath* the surface, aided by external impacts of solid bodies.

It was then a short time until water was in fact discovered underground, in the Moon's interior, in large quantities. [23] The water was actually discovered in lunar samples dating back to Apollo 17, and it was found in so-called "lunar melt inclusions," which were found to contain approximately 600 to 1, 400 parts per million of water. The quantity of water is so high that the interior of the Moon may contain as much water as the Earth's mantle. The melt inclusions consist of bits of magma that became trapped inside crystals before the magma as a whole was ejected during volcanic activity. The reason why the discovery of water even in the Moon was surprising is that the currently popular theory about the origin of the Moon predicts that the Moon should not have kept any water. The water found deep in the interior must have been present at the time the Moon formed. The problem is that the so-called "giant impact theory" that is believed by many to have created the Moon involves the Moon being formed in such a hot state that it was molten, and all volatiles (including water) should have been lost to space. The impact theory requires a large object, about the size of Mars, to have hit Earth, resulting in a massive ejection of material. That material is then supposed to have gone into orbit and eventually come together to form the Moon. Numerical computer simulations that attempt to model the process predict a variety of outcomes but the result of a single moon compatible with our own was argued to be feasible. [24] However, these simulations involve some hand-tuning and assumptions. Just because the simulations show that it is feasible it does not mean that the proposed giant impact actually took place and is the actual origin of the Moon. The finding of volatiles deep in the Moon's interior is obviously a big clue that it did not happen. An alternative scenario involves the Moon and Earth being formed simultaneously in a gas-giant "embryonic" structure (which loses its envelope to leave behind rocky cores). [25] However, this scenario is not without its caveats either and details have yet to be worked out. The problem of the origin of the Moon has again become open (the giant impact scenario was never very convincing anyway).

The Problem of Volatile Delivery

The discovery of substantial amounts of water on the Moon brings us nicely to an even more serious problem. That is the origin of water on Earth itself, and the general problem is referred to as the problem of the delivery of volatiles to rocky (terrestrial) planets such as Earth. The Earth could not have retained water and other volatiles while it was young and hot so it is generally thought that the Earth's water was delivered by comets and other solar system bodies. The problem can be studied quantitatively by measurements of the fractions of different isotopes of the constituents of water on Earth and in solar system objects. [26] However, such studies are ongoing and recent measurements have been used by a group of researchers to conclude that comets can only account for, at most, a few percent of Earth's water. [27] (The same study also concluded that some of the Earth's nitrogen may actually have been delivered by comets as well.) On the other hand, an asteroid covered in ice has been discovered, whereas it was previously thought that asteroids were too close to the Sun for ice to exist (i.e., they are inside the traditional snow line). [28] Clearly the origin of the Earth's water and the delivery mechanism is still an open question. The models are again complex and involve assumptions and uncertain factors.

Remember the discussion earlier in this book about how Jupiter can affect terrestrial planet formation, even though it is so far away? Well, it affects volatile delivery as well, and the effect of Jupiter has to be taken into account in the modeling.

Exoplanet Climatology and Habitable Zones

If we are going to ask what the requirements are for a planet to support life, the answers to this question are going to change rapidly in the coming years. It should be clear from the preceding discussions on extremophiles, and even the language used to describe them (including the name), along with our lack of understanding of the origin of life, that current ideas about the conditions for habitability are conservative and strongly centered around what we know of life on Earth (and what we know keeps expanding to include more and more previously unknown life-forms). Even the term extremophile screams of an inherent bias in what we think

171

the conditions for life *should* be like, given what *we* think are not extreme conditions. We do not ultimately know how life began, and if life can begin under many different conditions, then for those life-forms that can begin in extreme conditions, the term extreme is not at all appropriate. We don't even know if most of the biomass on Earth is actually beneath the surface. The issues and calculations involved in assessing habitability are very complex even when considering life-forms that we do know, and have been discussed in detail in the literature. Presumably, the reasons for theorizing about habitability in such exquisite detail is to target the most likely exoplanets that are habitable, in order to best allocate valuable observational resources. However, given the unexpected life-forms found on Earth in the most (previously thought) unlikely habitats and environments, it would make sense to maintain as broad a search as possible, and retain only loose criteria for the conditions of habitability. The value of performing excruciatingly detailed calculations of habitability, or worrying about effects that even life on Earth could withstand, is questionable. In the following sections I will give only a sketch of the current ideas without getting into too much unnecessary detail.

The Host Star, Tidal Locking, and System Stability

Conditions imposed on the host star as necessary for entertaining planets that can support life begin with the requirement that the star is in its stable hydrogen-burning phase (i.e., that the star is on the main sequence) for long enough for life to begin. For our Sun, the time on the main sequence is about 10 billion years and it is currently halfway through that phase. More and less massive stars have shorter and longer lifetimes respectively. Higher mass stars have higher surface temperatures. It is not known how much more massive or less massive than our Sun that a star should be before it is unable to support planets that can harbor life. It is premature to conclude that life-harboring planets could not exist around stars that have evolved off of the main sequence, on to the red giant, helium-burning phase. When the Sun reaches this phase its radius will expand by a factor of at least a couple of hundred, and it will become more luminous. Certainly, the inner planets will be engulfed, and the Earth will be fried. However, this phase goes on for more than a billion years and a scenario is possible in which a planet that was previously too cold because it was too far out from the host star, may well become amenable to supporting life after a star has left the main sequence.

Another thing that people worry about is "tidal locking." That's when the torques due to differential gravitational forces gradually slow down the spin of an orbiting body until it is forced to show the same face to the thing that it is orbiting. This has happened to our Moon. What people worry about is that if this happened to a planet, the "nightside" of the planet may be too frigid and inhospitable to life. However, it is once again worth remembering the amazingly diverse habitats of extremophiles on Earth. As for "higher" species, a population of aliens would simply be confined to the habitable zones on a tidally-locked planet, and the nightside may just be a holiday destination with artificial resorts, and a place where a handful of researchers live. Just think about the situation on Earth. We don't think it's strange that nobody lives at the North and South Poles, do we? Moreover, plenty of people have figured out how to live in Canada (I jest of course).

More worries come with the possibility of oblique orbits, eccentric orbits, or large tilt of a planet's spin axis relative to a line joining the planet and its host star (or any combination of these things). All of these irregularities would cause extreme climate variability and physical conditions. More will be said about these effects shortly. As for Earth, its orbit has a tolerable tilt, is virtually circular, and is nicely aligned with the plane perpendicular to the Sun's spin axis (i.e., it is not oblique).

Even if a planet has a nice orbit and a well-behaved host star, if the planet is a member of a multiplanet system, the possibility of dynamical instability arises. Although the gravitational force law is simple, the mutual interactions in a system with more than two constituents can result in very complex behavior that could in turn result in the ejection of a planet from the system, or at least a rough rollercoaster ride for a planet. People also worry about the variable illumination of the host star delivered to the planet, and/or that too little or too much illumination might be detrimental to life. Whether the concern is founded or not is of course unknown, considering that we know for a fact that there are life-forms on Earth that exist underground and in the deep ocean that don't need light. Any conclusions drawn about "higher" forms of life would be purely speculative.

Temperature and Albedo

Obviously the temperature of a planet is going to play a significant role in determining the habitability of a planet. Even considering the uncertainties about the temperature range for which life is possible, there are certain basic robust facts about various properties of matter that are inescapable. For example, at approximately 2,900 degrees Celsius (about 5,250 degrees Fahrenheit) and atmospheric pressure, about 10% of water molecules are dissociated into their components of hydrogen and oxygen. [29] The proportion of water molecules that are broken up continues to increase with temperature.

Estimating the temperature of a planet is an extremely complex business. For one thing, you know very well from your experience on Earth that a single number can't really represent the temperature of a planet because it's going to vary with place and time on the planet. You need a climate model that incorporates all of the physics, including the physical properties of the host star (luminosity, temperature), the planet-star distance, orbit and spin details, the planet's atmosphere, the planet's radiation *reflection* properties, the planet's surface properties such as land/ocean ratio (if relevant), the planet's chemical composition, and the planet's internal heat sources (if relevant). That's just a start. If the planet is a gas giant then the model must include as much of the physics of the entire planet as possible, not just of the surface (of course, in reality, many critical pieces of information will be missing).

It turns out that the reflective properties of a planet are very important for determining the energy balance and hence the planet temperature. The reradiation of the star's energy back into space depends on many factors including the composition of the surface and atmosphere (if present) of the planet, and even the presence of life on the planet can affect the energy balance. [30] A quantity called the *albedo* is used to characterize the radiation reflection properties of the planet. Since the albedo depends on the wavelength of the radiation, the planet's composition, structure, geometry, and other factors, there are many ways to define the albedo, which need not be a single number. Having said that, there is a particular definition of the albedo that *is* a single number, which represents a sort of average over wavelength and spatial structure, and from now on when I say albedo, I refer to this particular single number (technically speaking it is known as the *bond albedo*).

In general, it is not feasible to calculate the albedo from first principles, except for cases with very idealized assumptions (which may not be adequate). The albedo has to be measured by measuring the incident radiant energy and the reflected radiant energy. This can be done for Earth but for an exoplanet it is going to involve estimation based on what you think the composition and structure is, and on a comparison with known albedos measured locally. Therefore, attempting to estimate temperatures and climate properties of exoplanets is going to involve uncertainties due to assumptions about the albedo.

In lieu of full-blown climate calculations, it is possible to come up with a single temperature for a planet simply based on two things: the host star's luminous energy at the position of the planet, and the planet's albedo (usually an estimate). This temperature is known as the *effective temperature*. The word "effective" doesn't really have the usual meaning here. The physical assumptions behind this temperature are that the planet is naked (no atmosphere) and that it is in a steady equilibrium state in which it reflects back as much energy as it receives (if it didn't, it would not be in equilibrium, and the temperature would change until it did). Therefore, the effective temperature may have nothing to do with reality, but it is the temperature that you may see quoted in news reports or other articles. Most often, it is not possible to do any better than calculate only this effective temperature. Some of the effects of a planet having an atmosphere include distribution of energy over the planet (reducing temperature differentials), and heating due to greenhouse gases such as carbon dioxide and water vapor. We can see how much of a huge difference this can make from examples in our own solar system. The effective temperature of Venus computes to be -40 degrees Celsius (about -40 degrees Fahrenheit) but the actual surface temperature is 462 degrees Celsius (864 degrees Fahrenheit). This is a catastrophic "discrepancy" and is believed to be due to the greenhouse effect of the thick Venusian atmosphere. Moreover, without an atmosphere, the Earth itself would be a frozen snowball (but even *with* an atmosphere the physics is sufficiently complex that the Earth could *still* be frozen - see below). The effective temperature can be useful in the sense of providing a baseline number to work with, but the vast difference in temperature between Earth and Venus is important to bear in mind because both planets have similar masses, sizes, and distances to their host star.

Exoplanet Atmospheres

As we saw earlier when discussing close-in hot Jupiters, there are two competing effects that determine whether a gaseous planet is able to hold itself together. It is a competition between the self-gravity of the gas and the energy of motion of its constituent molecules and/or atoms, for which the temperature is a kind of "benchmark." The higher the temperature of the gas, the more likely it is that its constituents can fly apart against the self-gravity. The latter is greater the more mass that is compactified into a smaller region (but it does not quite have the same dependence on mass and radius as density). This is an oversimplified picture, and in reality more complex physical interactions between molecules/atoms have to be taken into account, as well as the effects of rotation, but that is the essence of it.

For planets that are fluid or solid as opposed to gaseous, there is an additional dominant effect, and that is the electrostatic force operating at the atomic/molecular level, and this works in the same direction as self-gravity, tending to help to hold the planet together. Internal thermal and mechanical activity may oppose this. Again, this description is oversimplified but captures the essential factors. In all cases (gaseous planets or otherwise) the gravity of nearby celestial objects will clearly affect possible outcomes.

A gaseous planet is of course *all* atmosphere in some sense (aside from a possible rocky core, but that is likely to be much smaller in size). I mentioned in chapter 2 that it has already become possible to study the gaseous composition of giant planets. However, for planets that are not predominantly gaseous, the question of whether a planet can hold on to a gaseous atmosphere is determined by the above principles (in essence). In other words if the thermal energy of gas on the surface of a solid or ocean planet is sufficiently large to overcome the surface gravity of the planet beneath it, the gas will escape into space. Since the thermal energy of a gas molecule or atom depends on its mass, not just the temperature of the gas, the composition of a gaseous atmosphere also affects whether it will be lost to space or not. Atmospheres composed of heavier elements are more likely to be held by a planet than those made of lighter elements. Hydrogen and helium atmospheres are therefore the first to go.

The observational study of exoplanet atmospheres is still a nascent and rapidly developing field. Since 2002, more than two dozen gas-giant exoplanet atmospheres have been studied. Most of the exoplanet atmospheric studies so far are on giant exoplanets because their size makes them more sensitive to measurements. Data taken in the infrared band have been particularly amenable to this (because of the higher planet-to-star contrast compared to visible light). One method utilizes so-called "secondary eclipses." The principle here is that because the atmosphere of a planet has a different "transparency" (or opacity) to the core of the planet (to the light from the host star), a distinct change in signal will be seen as the planet transits across the star, corresponding to the transition between the atmosphere and body of the planet doing the occulting. Studying the light with instruments that break it down by wavelength (a process known as *spectroscopy*), allows an observer to find the fingerprints of the atmosphere's compositional atoms and molecules (with many conditions and caveats of course). Already, sodium, water vapor, methane, and carbon dioxide have been detected in gas-giant exoplanet atmospheres. Even day-night temperature gradients and vertical atmospheric structure have been claimed to be measured. [31] A new technique involving the *polarization* of starlight reflected from a planet's atmosphere has been suggested, but the techniques still require further development, and results obtained so far are rather controversial. [32]

Direct imaging of exoplanets is highly desirable because it would yield more information than spectroscopy alone, but it is currently at the frontier of instrumental capability. Only a handful of examples exist at the moment, and those are exoplanets that are amongst the largest and brightest, and that are located at large distances from their host star. The latter property improves the planet-to-star brightness contrast. [33]

Studying the atmospheres of terrestrial exoplanets is much more challenging than it is for the gas giants, not only because the former are smaller, but because the atmospheric composition is expected to be much more diverse. Whereas the giant planets have retained their original atmospheres (which would have a composition similar to the host star), geological and biological activity can significantly change the atmospheric composition of terrestrial, rocky planets. Much of the original atmosphere could have been lost. The possibilities for the composition are so numerous that it is difficult to theorize what to expect. While no data are yet available for Earth analogs, a recent advance has been made with the observation of the super-Earth exoplanet, GJ 1214b (which has a mass of about 6 times that of Earth). The result was actually negative in the sense that no signatures of any elements or molecules were found, so it was only possible to rule out a restricted scenario, that of a cloud-free hydrogen atmosphere. [34] (In other words if the atmosphere *is* dominated by hydrogen, it must be cloudy.)

One of the ultimate goals in exoplanet atmospheric studies is to look for biosignatures, especially when Earth analogs are found. One of the strongest biosignatures of life on Earth is that due to oxygen. Even though oxygen is highly reactive, it makes up 21% of the Earth's atmosphere by volume. Without life, which is a continuous source of oxygen production on Earth, oxygen would be depleted. Another strong indicator of life is the so-called "vegetation red edge." This is a prominent feature at the red end of the reflection signal (spectrum) from Earth's atmosphere that is unique to plant life. [35]

Climate Solutions

To bring home just how complex, and sometimes nonintuitive, it is to determine the climate of a planet I will describe the results of some recent work on Earth's climate. In a paper published in 2007, Marotzke and Botzet reported the results of computer simulation experiments of Earth's climate in which they artificially turned off the Sun to see what would happen. [36] In other words, in their time-evolving simulations they manually switched off the solar irradiance. The idea was to emulate (albeit very crudely) the result of a catastrophic event that might blast a lot of dust or other contaminant into the atmosphere and block out the Sun. Such an event could be due to, for example, a giant asteroid impact, or a massive volcanic eruption. What the researchers found is that for the given Earth-Sun distance and current carbon dioxide level (i.e., they did not adjust either of these), *two very different stable solutions exist.* The two solutions are very different from each other: one is a completely frozen Earth (a "snowball" Earth), and the other is a warm Earth like the one we know now. People generally do not appreciate that the physics of a complex system such as Earth can lead to very different solutions for the same physical parameters. Most people would be terribly confused and would wonder how this fits in with global warming. Is there an intuitive way to understand this? Not easily in the brief amount of space that I'm going to allocate to this topic. However, as a kind of "demonstrative statement," think about the fact that a thick sheet of ice, although very cold, will act as an insulating layer for any heat below it. Ice also has very different reflective properties to the regular warm matter on Earth. The overall temperature of anything is determined by the balance of heating and cooling.

The details of the dual solutions found were also rather interesting. After the solar irradiance was set to zero in the simulations, it took about 15 years for the Earth to become completely covered with ice (including the oceans freezing). However, the researchers also found that there is a hysteresis effect in the sense that when the solar irradiance was turned back on after the freezing up of Earth, the ice did not go away (at least for the duration of the numerical simulations). In other words, deglaciation does not automatically follow when the Sun is turned back on. The conclusion from this is that a "brief" dark spell, such as that which would result from an asteroid impact or a large volcanic eruption, could plunge the Earth into a prolonged snowball state.

The Faint Young Sun Problem

Here is another unsolved problem related to Earth's climate. By inference, it constitutes a source of uncertainty in understanding exoplanet systems because if there is something wrong with our

181

understanding of the Earth-Sun system, whatever it is that is wrong may carry over to exoplanet systems. The problem is that stellar evolution models predict that the Sun was fainter when the Earth was much younger, but geological evidence of Earth's climate at that time contradicts this. Specifically, 2.3 billion years ago, the Earth's water should have been frozen due to the Sun being 30% dimmer than it is now, but geological evidence from liquid water-related sediments dated at 3.8 billion years, and evidence of early life-forms dating back to 3.5 billion years, conflict with this. [37] The problem is usually referred to as the "the faint young Sun paradox." However, the word "paradox" in scientific discourse is usually a euphemism for "problem," implying a grave *error*. Yes, there is a mistake somewhere along the line and nobody has figured out what the mistake is. (Incidentally, I like the *Merriam-Webster (unabridged)* definition of euphemism in this context: "a polite, tactful, or less explicit term used to avoid the direct naming of an unpleasant, painful, or frightening reality.")

What went wrong? Aspects of stellar evolution theory could be wrong. The theory gives a prescription of how the luminosity of a star increases as it progress in its hydrogen-burning phase (on the main sequence). Certain assumptions go into that. One of the factors that affects the prescription is loss of mass (i.e., mass ejection). Some have argued against mass loss as a resolution of the problem but ultimately we don't know what the mass loss rate of the Sun was billions of years ago. An alternative explanation is that there was a much stronger greenhouse effect on Earth due to the composition of the atmosphere being different to what we think it was. Claims to have solved the problem, and counterclaims continue. A recent alternative put forward in 2010 based on making the young Earth's clouds and surface less reflective (giving them a different albedo) was shown in 2011 to not resolve the problem. [38]

Circumstellar Habitable Zone

Finally, we come to talking about the standard "circumstellar habitable zone," or CHZ. This is a concept that was invented before exoplanets were actually discovered, in order to quantify the properties of a star-planet system that were amenable to supporting life. It is based on the premise that life *requires* liquid water and the CHZ is a spatial region corresponding to distances from the host star at which a planet can support liquid water. [39] Although water has some remarkable properties, [40] including the wide temperature range over which it is liquid and its ability to act as a solvent in a wide-range of circumstances, we do not ultimately know if the premise is true. In fact the continuing discovery of extremophiles do not support the premise. The necessity of liquid water (and therefore the restrictiveness of the CHZ) has begun to be questioned in the literature. Bacteria have been discovered in the South Pole that can reproduce at −10 degrees Celsius (14 degrees Fahrenheit) and that can exist at −85 degrees Celsius (about −120 degrees Fahrenheit). Strains of *archaea* have been cultured that remain viable at temperatures higher than the boiling point of water, up to about 130 degrees Celsius (or about 270 degrees Fahrenheit). [41] Still, if we are talking about "higher," intelligent life-forms, you might argue that liquid water is definitely needed in that case. Ultimately, we don't know.

Also implicit in discussions of habitable zones is that we are talking about terrestrial rocky planets. Gas-giant planets are generally dismissed for potentially harboring life. Again, we ultimately don't know if the rocky core in gas giants (or even ice giants) could harbor life. Moreover, our own solar system has several satellites (moons) of the gas-giant planets that could harbor life, and those satellites are outside the conventional habitable zone for our Sun.

Even if we accept the liquid water criterion, it should be clear from the preceding discussions that calculation of the habitable zone is going to be riddled with complications. The list of issues is very long and I will only mention a few here. The freezing and boiling points of water depend on the pressure, so there is going to be some uncertainty from this. Many complexities of climate calculations have already been discussed. These all involve a large number of assumptions and adjustable parameters. Whatever the detailed assumptions are, two more complications are that the CHZ depends on the properties of the host star as well as the planet, and that the CHZ varies with time as the host star evolves, and as the planet's properties change.

A recent study that criticized the standard definition of the CHZ evaluated the so-called "fractional habitability" of Earth as a function of seasonal effects, planet rotation rates, and land to ocean ratios. [42] The fractional habitability quantifies the fraction of a planet that is habitable (under a given set of assumptions). The study found that the predicted fractional habitability for Earth is different to its actual fractional habitability that we know. Another study by the same researchers investigated the effects of the obliquity of the orbit and eccentricity of the orbit on a planet's habitability and found the effects to be important. [43]

Although in principle the inner and outer edges of the CHZ are straightforward to define, calculations in the literature differ in the assumptions and approximations involved in the complex physics with the outcome that different researchers can come to different conclusions about the habitability of the same planet. The inner edge of the CHZ (closest to the host star) is determined by the breakup of water into its constituents due to intense radiation. The outer edge of the CHZ (farthest from the host star) is affected by the condensation of carbon dioxide out of the atmosphere, which affects the reflection (scattering) of light, and therefore the energy balance (and therefore the temperature). The so-called carbon-silicate cycle affects the outer edge of the CHZ. This cycle is complex and has a very long timescale (millions of years) and involves the reaction of carbon dioxide (via rain, making carbonic acid) with silicates in rock to make carbonates which are incorporated into living creatures. The carbonates are eventually returned to Earth, after which volcanism then returns carbon dioxide to the atmosphere. This is of course a grossly oversimplified picture, but the main point is that there are a large number of uncertainties in implementing the physics to calculate the CHZ. A study of the multiplanet system Gliese 581 in 2011 summarized this sentiment nicely: "The precise inner and outer limits of the climatic habitable zone are still unknown owing to the limitations of the existing climate models." [44] The bottom line is that whenever you read about an exoplanet being in the habitable zone, you should be aware that at the moment the conclusion is likely not to be definitive and that the claim is open to debate. Also note that for the *Exoplanets App* (http://exoplanet.hanno-rein.de/), which nicely shows visualizations of some attributes of cataloged exoplanets, the habitable zone boundaries are based on just one particular formulation (and therefore one particular set of assumptions). [45] So those habitable zone boundaries are not definitive. Hereafter, when I say "the" habitable zone, I mean *a* habitable zone that corresponds to the particular definition that has been defined and adopted by the researchers and/or study that is mentioned.

The system Gliese 581 is interesting because it is a multiplanet system that has received a lot of attention, and it has been well studied because it harbors more than one super-Earth that is potentially in the habitable zone. [46] The Gliese 581 system has four definite planets and two additional planets whose existence is not certain (because they are inferred to exist indirectly from solutions of dynamical equations). Of the planets whose existence is certain, one (Gliese 581c) was originally thought to be at the inner edge of the habitable zone, and another (Gliese 581d) was thought to be located at the outer edge of the habitable zone. However, detailed calculations have shown that the exoplanet Gliese 581c is too close to its host star to maintain liquid water on its surface and would incur a runaway greenhouse effect unless a specific cloud distribution is invoked. [47] On the other hand, Gliese 581d has been claimed to be the first rocky super-Earth that is potentially habitable. The minimum mass of Gliese 581d is about 7 Earth masses. In a recent paper on Gliese 581d, the claim for habitability is tempered by a statement in the paper: "Future observations of atmospheric features can be used to examine if our concept of habitability and its dependence on the carbonate-silicate cycle is correct, and assess whether Gliese 581d is indeed a habitable super-Earth."

In the paper describing the February 2011 release of *Kepler* data, 54 planet *candidates* were listed as "in the habitable zone". [48] However, this was determined using only the simple effective temperature, and an assumed nominal albedo. In addition to the fact that the objects are not confirmed planets, the caveats discussed above, plus significant uncertainties about some of the host-star properties, obviously make the list very tentative.

Galactic Habitable Zone

In addition to the habitability zone around a star (CHZ), a habitability zone for our galaxy has also been suggested (uninspiringly called the "Galactic Habitable Zone," or GHZ). The basic idea behind the concept is that certain times and positions within our galaxy (the Milky Way) are favorable to the existence of Earthlike planets and life. Factors that may affect the GHZ are, for example, the relative abundances of heavy metals (i.e., elements heavier than Helium) as a function of position and time in a galaxy, and the proximity of potentially habitable planets to dangerous cosmic explosions (such as supernovae, stars that explode after a certain stage in their evolution). However, the problem is that the factors that affect the GHZ are not well understood and none of the relevant factors can be quantified in a meaningful way. Indeed, the GHZ may be yet another unnecessary level of "worry," given that we don't really understand the exact conditions necessary for life anyway. In a recent paper that questions the usefulness of the GHZ, the author (N. Prantzos), performed some calculations on various constraints but had to make many assumptions that he said, "...are far from being well founded at present." The author concludes: "Thus, the concept of a GHZ may have little or no significance at all." [49]

You may come across the term "Goldilocks zone," or "Goldilocks planet." These are just further descriptions expressing the habitability of a planet in terms of a multitude of factors being "just right" (fine-tuned) for supporting life. As you might have guessed, I don't like the "Goldilocks" terminology either. (What exactly does "just right" mean in this context?)

Is There a New Answer to the Old Question?

We now consider whether all of the amazing new discoveries in recent years shed any new light on the old question, "Are we alone?" With respect to the question of whether Earth (and the existence of life) is very special, there are two schools of thought that could not be more different from each other. One is the so-called "rare Earth hypothesis," which is exactly what its name suggests. [50] In this scenario Earth's physical conditions and intelligent life on Earth are the result of a very large number of very unlikely events, with the result that the probably of finding another example like it is in essence zero. The other view is the opposite of that, supported by the fact that on Earth we continually keep finding life in the most unlikely and unexpected places. [51] In this scenario there would be undiscovered principles that lead to inevitable structures that are living organisms. It is possible to argue both ways, and people have written entire books making the case for one or the other. The latter fact alone suggests the simple conclusion that nobody has absolutely any idea which one is closest to the truth.

Every book that I have come across that touches on the "Are we alone?" question discusses the so-called "Drake equation." The "equation" is supposed to give an estimate of the number of "communicating" civilizations in our galaxy. I am not going to discuss the "Drake equation" here because, as the same books admit, almost every term in the "equation" is completely unknown, so you can come up with any answer you like. You can read about the Drake "equation" elsewhere if you are interested. Sorry to break the bad news, but nobody has any idea of the answer to the age-old question. I will say, however, that the term in the Drake "equation" that corresponds to the "length of time that intelligent civilizations send out signals for other aliens to receive" is really a financial term, and the financial reality is rarely discussed. How much money is a species prepared to put into transmitting signals into space, as a planet-wide effort? On Earth, the planet-wide financial commitment to send signals into space for aliens to receive is exactly zero (making the rest of the Drake "equation" irrelevant). Whilst millions of individuals are prepared to make a financial commitment to send signals to aliens, for humans as a species the activity has zero priority. The number-one planet-wide financial commitment is for humans to make weapons designed to exterminate other members of the species. These are facts. If other alien species exist, and their financial priorities are the same as humans, then the chance of us finding them is exactly zero. The various incarnations of SETI (search for extraterrestrial intelligence) on Earth have received limited support of various kinds, and you can read about the various activities and financial woes on a variety of websites. [52] Still, it is true to say that there has *never* been a project that vows to continuously pump signals at astronomically "serious" power levels into space. Also, it is important to remember that SETI-type activities predominantly aim to *receive* signals, not to send them. If we are not sending out signals, why should we expect an alien species to do so? It would be very hard for any species to inject serious money into something that will likely have no result for generations, and certainly no financial return.

There is one factor that is becoming more tractable as our knowledge about exoplanets increases. That is, we can begin to estimate the fraction of sunlike stars that have Earthlike planets. What is needed for this is a systematic survey that is corrected for observational biases and selection effects (it cannot be done for the hundreds of exoplanets confirmed so far because they were discovered heterogeneously). This has been attempted for the approximately $150,000$ stars observed by the *Kepler* mission, along with the exoplanet *candidates* that have been found. The candidates need to be confirmed as exoplanets. Nevertheless, it was found in a study in 2011 that the occurrence of habitable planets in the *Kepler* sample is $1.4\pm0.5\%$, using a particular set of assumptions about the habitable zone. The same study found that using a habitable zone based on different (more optimistic) assumptions gave an answer of $2.7\pm0.5\%$. [53] Needless to say, there are many caveats and assumptions that go into producing these numbers, but prior to dedicated planet-finding missions, it was not possible to even attempt such estimates. The estimates can be continuously refined with more and more follow-up work. Overall, the prediction is that 12 Earth analogs will be found by the time that the *Kepler* mission ends, with the number found so far being 4 (but these need to confirmed). More specifically, 4 candidates in the Earth to super-Earth range have been found in the *Kepler* data, and these are estimated to have effective temperatures below 27 degrees Celsius (or about 80 degrees Fahrenheit, or 300 degrees on the "Kelvin" scale).

I will now use the numbers above, as preliminary as they may be, to take a wild guess at how many habitable planets there may be in our galaxy, the Milky Way (according to the same criteria in the literature reporting the results from the *Kepler* survey). The conditions for habitability are fairly conservative and we can be even more conservative by applying the lower error margin given on the occurrence of habitable planets in the *Kepler* sample (i.e., we take $1.4 - 0.5 = 0.9\%$ as the percentage of habitable planets in the survey sample). Now, $1,235$ exoplanet candidates out of $150,000$ stars examined, of which 0.9% are habitable, amounts to a hit rate of 0.0074%. The Milky Way has on the order of 100 billion stars, so this translates to 7.4 million habitable planets. Even if our estimate is off by a factor of $1,000$, that's still $7,400$ habitable planets. The numbers are quite staggering. However, our galaxy is approximately $100,000$ light years from one end of its disk to the other (and about $30,000$ light years wide in the shortest dimension). As we shall see in the next section, it is going to be hard enough for humans to figure out how to get to even the closest exoplanet (that is about 10 light years away).

Even further from our reach are other galaxies. Galaxies are typically separated from each other by millions of light years (the closest to us is Andromeda, located at a distance of about 2 million light years from Earth). The number of galaxies in the Universe can only be estimated but it is thought that there are at least a hundred billion in the observable Universe (each galaxy contains on the order of 100 billion stars). This would then put the number of habitable planets in the Universe at more than 0.7 billion billion for the least conservative estimate, and at 0.7 million billion for the more conservative estimate. This means that if there are at least two planets in the Universe that harbor intelligent life (including Earth), then the probability of a habitable planet resulting in intelligent life need not be higher than 3 parts in a million billion. However, this estimate is actually not very meaningful because there is no such thing as a "now" that applies to all parts of the Universe since simultaneity is not applicable in the usual way, and we have not considered how long a civilization needs to exist for. All of these things (and more) are difficult even to take wild guesses at.

How Long Would It Take to Get to the Nearest Exoplanet?

What are the prospects of actually traveling to a planet outside our solar system? I mentioned that currently the distance between us and the confirmed exoplanets (in the October 2011 sample) lies in the range of just over 10 light years to just under 28,000 light years. Given the physical limit of the speed of light that means that the absolute minimum travel time (in the reference frame of the traveler) is just over 10 years, but achieving velocities close to that of light is way beyond our means. Velocities that are currently achievable are more like 20,000 miles per hour (or about 32,000 kilometers per hour) and at that speed it would take about 335,000 years to cover a distance of 10 light years (yes, that's a third of a

191

million years). A velocity of 20,000 miles per hour is less than 0.003% of the speed of light. Of course we can expect continued improvements in our capabilities to achieve higher speeds but an exponentially increasing amount of energy is required to approach the speed of light. At this time it is difficult to envisage how velocities of 99% of the speed of light or more will be achievable. If it is ever possible, the so-called "time dilation" effect of relativity comes into play, which leads to some interesting consequences. At 99% of the speed of light the frame of reference that the traveler left behind (home) and the parts of the rest of the Universe that had velocities relative to home that were much less than that of light, age approximately seven times slower than the traveler. Thus, if the traveler made a round trip that took 30 years (not stopping at the destination but just turning around), upon return, the traveler's home planet and everything on it will have aged by about 210 years. So although the time dilation is beneficial for the traveler, it is bad news for anyone back home who wants to know what happened. For exoplanets that are further away, millennia could pass before the home planet witnesses the return of the traveler, or receives signals from the traveler upon landing on the alien planet. Such experiments and programs would have to be planned for durations of hundreds to thousands of years across multiple generations. This is truly mindboggling.

We are near the end of the book. I'm sorry if you're disappointed that there have been no answers in this chapter, only questions. This is not a depressing state of affairs, but rather it is actually very exciting that we know so little because it means that the arena is wide open for learning things that are completely new. If you are young and still at school, and have ambitions to become a space scientist, this is a great time for you. Don't believe anything your textbook tells you without thoroughly questioning it. Time is on your side because by the time you graduate, the old fogies will still be trying to fit a square peg into a round hole, stubbornly unable to let go of ideas and paradigms that don't work. Roll your sleeves up, there is work to be done! There are profound discoveries to be made, and they are waiting for you.

Ant School

I will leave you with a final thought. Suppose that you are an ant. You are sitting in class and teacher ant is explaining how antkind is vulnerable to a particular kind of catastrophic event. During these events, first the ground shakes, accompanied by loud and terrifying thumps. Then the sky darkens. The darkening increases, along with a giant elliptical object falling out of the sky and obliterating any ants that could not get away in time. The phenomenon, teacher ant explains, is called *feetsquish*. Teacher ant explains that ant scientists have tried to understand and tried to predict *feetsquish* for hundreds of years without success. There has been some evidence to suggest a correlation in the sense that *feetsquish* is less common under rainy conditions. However different studies have produced conflicting results and *feetsquish* remains unpredictable with respect to when and where it occurs.

In reality, the ants live side-by-side with an intelligence that is far, far superior to them. But how could the ants possibly know that (being, presumably, far less intelligent)? Yet that superintelligence is living "right under their ant noses." To the ants, being killed by humans stepping on them would be classified as a phenomenon that is just part of nature, being as unpredictable as some other natural phenomena such as lightning. The ants lack the intelligence to figure out that there is a superior intelligence living in parallel with them that is in a completely different league.

Not only is it misguided to suppose that we, as humans, occupy the highest possible position in intelligence, and that there is no way that we would *not* recognize a much higher intelligence, there is no evidence for it. Consider also that not only are ants unable to communicate with the higher intelligence, the higher intelligence cannot communicate with the ants. You cannot tell the ants to leave your kitchen counter alone. They do not have the apparatus to understand your request. Why should we necessarily expect to be able to recognize and communicate with *much* higher forms of intelligence than ourselves, even if we lived side-by-side with them?

Appendix A

What Is a Planet?

You would think that it would be a fairly simple matter to define a planet as a "round thing that orbits a star." However, the situation is actually very messy, partly because it has become the domain of committees, and partly because a planet is, in the end, a rather arbitrary concept. It also turns out that planets in our solar system and outside our solar system (exoplanets) are handled differently at the moment, mainly because observational information on the exoplanets and their star systems is currently far less detailed than it is for our own planets. For example, as you will see below, the official definition of a planet requires you to know whether an object is "nearly round," but exoplanets are so far away and have an apparent size that is too small to be able to determine "roundness," in general. In addition there is a class of objects that does not orbit a star (free-floating planets) that remains formally undefined. Such objects have been claimed to be detected, and even claimed to be more common than stars in our galaxy. [1] From both an observational and theoretical view, things are rather up in the air with respect to how a planet outside our solar system should be defined, and whether free-floating objects should even qualify to be evaluated for "planethood."

With respect to planets in our own solar system, the official "resolutions 5 and 6" of the 26th assembly of the *International Astronomical Union* (IAU) set out the formal definition, and how Pluto should be categorized. One of the problems with coming up with a definition is concerned with how to treat large satellites (moons). The official definition put forward by the IAU can be summarized as follows: A planet is an object that (1) orbits the Sun, (2) is massive enough for its own gravity to make it nearly round, and (3) has cleared its orbit of debris. The definition explicitly excludes satellites, but as we shall see, there is a grey area here. Objects that satisfy the first two of these conditions and not the third are to be called *dwarf planets*, and Pluto is officially classified by the IAU as a dwarf planet. The shockingly vague and unscientific language of the IAU definition ("nearly round," "cleared its orbit") is still rather controversial.

Now, the satellites of the eight solar system planets are safely nonplanets and they are not dwarf planets either, and the IAU says this is because the satellites are so much less massive than their parent planets that the so-called center of gravity of the system is inside the parent planet. However, there is a problem with *Charon*, which is associated with Pluto (traditionally as its moon) and is comparable in size to Pluto. The center of gravity of the system is in free space and not inside either of the two objects so they look like they are orbiting a point in free space. It is important to realize that *both* Charon and Pluto are still orbiting the Sun at the same time. That means that *Charon* orbits the Sun but strictly speaking does *not* orbit Pluto, even though it is gravitationally bound to Pluto. There is nothing in the IAU definition of a planet or a dwarf planet that dictates where the center of gravity between two solar system bodies should be in order to violate the first condition in the definition of a planet, so what exactly constitutes a satellite is ill-defined. The position of the IAU is that, "*For now, Charon is considered just to be Pluto's satellite. The idea that Charon might qualify to be called a dwarf planet in its own right may be considered later.*" What "later" means is of course anybody's guess.

So how many dwarf planets in our solar system are there so far? Officially there are five: *Ceres (the asteroid), Pluto, Haumea, Eris, and Makemake*. However, there are many more candidate dwarf planets amongst the many rocky/icy objects beyond Neptune and Pluto, in the so-called Kuiper Belt, located at roughly 30 to 50 AU from the Sun. However, for the time being the IAU accepts only five as official dwarf planets. The other objects are known as KBOs (Kuiper Belt Objects), or TNOs (Trans-Neptunian Objects). Their relatively small sizes are very difficult to measure. To complicate things further, the term *plutino* is used to describe objects in the inner Kuiper Belt that follow similar orbits to Pluto in the sense that the orbital period is locked in resonance with Neptune's orbital period (2 plutino orbits for every 3 orbits of Neptune). Pluto is a plutino. Got that? Now, to make your head spin even faster, a *plutoid* is a TNO that qualifies for being a dwarf planet. This means that Ceres is a dwarf planet but not a plutoid (because it is not a TNO). Plutoids are too small to actually tell whether they are "nearly round" so the judgment is made on the basis of a theoretical guess. A plutoid is distinguished from a plutino in that it does not have the orbital resonance with Neptune that a plutino does. Pluto is a plutoid. Satellites of plutoids are not themselves classed as plutoids (refer to the discussion about Charon for the obvious problems with this). The reader is referred to the solar system review in Chapter 2 for more details on the Kuiper Belt and beyond. You can find more details on the dwarf planets and keep up with the latest findings at another IAU website (http://planetarynames.wr.usgs.gov/Page/Planets/) that also includes details of the satellites of the dwarf planets, as well as details of the eight "classical" planets and their satellites.

Finally (as far as the solar system is concerned), what are "small solar system bodies"? Basically the IAU defines this term as the official designation of anything that misses the definition of a being a planet, a dwarf planet, or a satellite. This includes asteroids and comets.

Does the Classification of a Solar System Object Depend on How It Was Formed?

No: the IAU explicitly did not include a "formation criterion" because there are too many theoretical uncertainties. In other words, formation of solar system objects is not yet sufficiently understood (see chapter 4).

What Are the Mass Limits for a Planet?

The IAU definition of a planet also does not specify an upper boundary on the mass of an object. However, it is known that if the mass of an object exceeds about 12 to 13 times the mass of Jupiter, it may begin to burn what is known as *deuterium* (hydrogen with two neutrons per atom instead of one), and such nuclear burning would then classify the object as a *brown dwarf*. [2] The exact "critical" mass is subject to theoretical uncertainty because the calculations involve uncertain parameters and assumptions. Objects that exceed the critical mass are known as brown dwarfs, which are essentially extremely dim stars. Note that nuclear burning of deuterium occurs before hydrogen because the temperature required for the ignition of nuclear burning is lower for deuterium than it is for hydrogen, even though deuterium is far less abundant than hydrogen.

What about the lower end of the mass range? Is there a minimum mass that can qualify as a planet? This is essentially determined by the "round" condition in the definition of a planet. Objects that are held together by their own self-gravity are prevented from collapsing completely by various sources of opposing pressure. In the absence of internal heat sources, for objects that have densities that are typical of planets, the principal sources of resistance are the electrostatic forces in the atomic and molecular structure. Every part of a self-gravitating structure is always trying to fall and collapse to the center. If this self-gravitational force is sufficient to overcome the resistive forces then the matter falls radially until new resistive forces increase enough to prevent further collapse, and opposing forces balance each other again. Since the gravitational force is always directed towards a single "point," the center of mass, the resulting equilibrium structure will tend towards being spherical. However, if the object was not massive enough and compact enough to begin with, gravity will not be able to crush and redistribute the matter, and the equilibrium configuration will not be spherical. In general, estimating the critical mass and radius for an object to become spherical is complex. It depends on the detailed composition and the physical state of the matter (for example, whether the object is rocky or not, and whether the rock was molten at the time of formation). The asteroids provide good examples: the smaller asteroids have no particular shape, whilst some of the most massive ones (including Ceres) are spherical. For objects that are further out in the solar system, which are typically made predominantly from ices, a critical size of about 400 kilometers wide has been estimated. [3] Although the physical principles that determine the lower limit of the mass of a planet are straightforward, in practice it is not simple to calculate that lower limit.

The next question is, what is the maximum mass that a brown dwarf can have before it becomes a fully-fledged star, burning regular hydrogen in nuclear fusion reactions? The question is important because if a candidate exoplanet has mass measurement uncertainties that straddle the planet/brown dwarf boundary (13 Jupiter masses), we would like to know whether we can rule out the object being a star. To become a fully-fledged star, an object must be massive enough for the pressure from nuclear burning to balance the tendency to collapse under its own gravity. In other words, the object is not supported by the intermolecular electrostatic forces, nor by nonnuclear thermal pressure, against gravitational collapse. Calculations show that the critical minimum mass for a star is about 80 times the mass of Jupiter (but the exact value depends on the chemical composition and other factors). [4]

So What Is an Exoplanet?

We see from the above discussion that the first condition that must be met for an object to qualify as a planet is that its mass must be less than the critical mass at which thermonuclear burning of deuterium can begin. At the lower mass end, the definition of a planet is consistent with that for our solar system: the object must be massive and compact enough that its self-gravity makes it "nearly round." The condition of "clearing the orbit of debris" also carries over. However, in addition, there is the issue of whether an object that is not associated with a star (i.e., a "free-floating" object) should be called a planet (even if it has a mass of less than 12 to 13 Jupiter masses). For the moment, such objects are not labeled as planets, but are labeled as "subbrown dwarfs." However, sometimes they are referred to as "planemos" (short for "planetary mass objects").

In practice, it may not be possible to measure the required observables for an exoplanet to unambiguously determine its formal qualification as a planet. In particular, only a small percentage of planets have yet been imaged, and it is not possible, in general, to say with confidence that an object is "nearly round," and the issue really begs a more quantitative notion of what exactly is meant by "nearly round" anyway. There is a document that was produced by the "IAU Exoplanet Working Group" in 2006 (http://www.dtm.ciw.edu/boss/IAU/div3/wgesp/) that describes the ins and outs of the classification of exoplanets, and it is still quite an ongoing affair.

Appendix B

Selected Websites

The Extrasolar Planets Encyclopedia
URL: http://exoplanet.eu/
This website hosts many technical resources, including up-to-date tables of numerous exoplanet parameters, as well as comprehensive documentation of source research papers.

Kepler mission
URL: http://kepler.nasa.gov/
This is the official website of the NASA *Kepler* mission. The website provides access to some real data and some educational resources, including some presentations.

CoRoT Mission
URL: http://smsc.cnes.fr/COROT/
The website for the *CoRoT* (*Co*nvection, *Ro*tation, and *T*ransits) Mission. The website has a variety of background information, news, and presentations.

Exoplanets Data Explorer

URL: `http://exoplanets.org/`

This website has a powerful tool that allows the user to analyze and plot data, suitable for those with some appropriate knowledge and expertise in data analysis.

Exoplanetology

URL:
`http://s3.amazonaws.com/exoplanetology/index.html`
This website has news and articles on exoplanets, some tools and links, and a blog.

Planet Quest

URL: `http://planetquest.jpl.nasa.gov/`
The official NASA JPL (Jet Propulsion Laboratory) website for exoplanets. There are educational resources for both students and teachers.

Planet Hunters

URL: `http://www.planethunters.org/`
A website for those who wish to contribute and help to find exoplanets from real data. There is also a blog.

Transit Search

URL: `http://transitsearch.org/`
A website for serious amateur astronomers interested in observing transiting exoplanets.

Exoplanets App

URL: `http://exoplanet.hanno-rein.de/iphone/`

A free *App* for *iPhone*, *iPad*, and *iPod touch* that gives access to an exoplanet database that is updated daily and includes interactive visualizations. Note that the habitable zone calculated by the *App* adopts a particular definition (that of F. Selsis, et al., "Habitable Planets Around the Star Gliese 581?" *Astronomy and Astrophysics* 476 (2007), 1373-1387), and should not be regarded as definitive because there is *no* definitive definition of the precise boundaries of the habitable zone.

Exoplanets Digest

URL: `http://exoplanetsdigest.com/`

A blog by the author of *Exoplanets and Alien Solar Systems* that stands above the noise of exoplanet news, providing "translations of astrophysics" from selected real research papers into plain language, without the razzamatazz that often obscures (or even changes the meaning of) the actual results.

SETI Institute

URL: `http://seti.org/`

This is the website of an organization dedicated to SETI (Search for Extraterrestrial Intelligence).

International Astronomical Union (IAU)

URL: `http://iau.org/`

The official website of the IAU.

NASA solar system website

URL: http://solarsystem.nasa.gov/

This website has detailed information on solar system planets and objects, as well as various NASA missions.

JPL solar system pages

URL: http://www.jpl.nasa.gov/solar-system/index.cfm

This is the NASA *Jet Propulsion Laboratory* (JPL) website and has resources and information on the solar system.

Notes

Chapter 1

1. (*Chapter 1*)
B. McKernan and T. Yaqoob, "Occultation Mapping of the Central Engine in the Active Galaxy MCG -6-30-15," *The Astrophysical Journal* 501 (1998), L29-L33.

2. (*Chapter 1*)
My principal research activities have been predominantly in the areas of X-ray astronomy, active galaxies, quasars, and the high-energy environment and physical processes in candidate black-hole systems.

3. (*Chapter 1*)
Distances between Earth and Jupiter were calculated using the tool at http://ssd.jpl.nasa.gov/horizons.cgi (NASA Jet Propulsion Laboratory, "HORIZONS system" ephemerides generator). The "current distance" between Earth and Jupiter that was used was 5.94 AU. The distance between Earth and Jupiter varies between 4.21 AU and 6.44 AU (approximately). Planetary mass data from http://solarsystem.nasa.gov/planets/ were used. The distances between the centers of mass of mother and baby were as-

sumed to be 0.1 and 0.5 meters for the close-hug and arm's length holding positions respectively.

Chapter 2

1. (*Chapter 2*)
See, for example, http://solarsystem.nasa.gov/planets/ for more precise planetary data.

2. (*Chapter 2*)
J. S. Greaves, Helling, Ch., and Friberg, P., "Discovery of Carbon Monoxide in the Upper Atmosphere of Pluto," *Monthly Notices of the Astronomical Society* 414 (2011), L36-L40.

3. (*Chapter 2*)
See, for example, NASA solar system website for more quantitative data on dwarf planets.

4. (*Chapter 2*)
M. E. Brown, et al., "2002 AW197," *Minor Planet Electronic Circ.* 2002-O30 (2002). See also, M. E. Brown and C. A. Trujillo, "Direct Measurement of the Size of the Large Kuiper Belt Object (50000) Quaoar," *The Astronomical Journal* 127 (2004), 2413-2417; W. C. Fraser and M. E. Brown, "Quaoar: A Rock in the Kuiper Belt," *The Astrophysical Journal* 714 (2010), 1547-1550. The latter paper also includes results for a moon of Quaoar (named Weywot).

5. (*Chapter 2*)
The outer edge of the Kuiper Belt is inferred from a drop-off in Kuiper Belt Objects after analysis of survey data (incorporating corrections for observational bias), and modeling of those data.

See for example: D. Jewitt, J. Luu, and C. A. Trujillo, "Large Kuiper Belt Objects: The Mauna Kea 8K CCD Survey," *The Astrophysical Journal* 115 (1998), 2125-2135; R. L. Allen, G. M. Bernstein, and R. Malhotra, "The Edge of the Solar System," *The Astrophysical Journal* 549 (2001), L241-L244; J.-M. Petit, et al., "The Kuiper Belt Luminosity Function from $m_R = 22$ to 25," *Monthly Notices of the Astronomical Society* 365 (2006), 429-438.

6. (*Chapter 2*)
M. E. Brown, C. A. Trujillo, and D. Rabinowitz, "Discovery of a Candidate Inner Oort Cloud Planetoid," *The Astrophysical Journal* 617 (2004), 645-649.

7. (*Chapter 2*)
M. Schwamb, M. E. Brown, and D. Rabinowitz, "A Search for Distant Solar System Bodies in the Region of Sedna," *The Astrophysical Journal* 694 (2009), L45-48.

8. (*Chapter 2*)
J. J. Matese and D. P. Whitmire, "Persistent Evidence of a Jovian Mass Solar Companion in the Oort Cloud," *Icarus* 211 (2011), 926-938, and references therein.

9. (*Chapter 2*)
See *CoRoT* website. See also, S. C. Maciel, Y. F. Osorio, and J. R. de Medeiros, "Ten *CoRoT* Eclipsing Binaries: Photometric Solutions," *New Astronomy* 16 (2010), 68-71.

10. (*Chapter 2*)
The discovery of sodium in HD 209458: D. Charbonneau, T. M. Brown, R. W. Noyles, and R. L. Gilliland, "Detection of an Ex-

trasolar Planet Atmosphere," *Astrophysical Journal* 568 (2002), 377-384. The Discovery of carbon and oxygen in HD 209458: A. Vidal-Madjar, et al., "Detection of Oxygen and Carbon in the Hydrodynamically Escaping Atmosphere of the Extrasolar Planet HD 209358b," *Astrophysical Journal* 604 (2004), L69-L72.

11. (*Chapter 2*)
G. F. Benedict, et al., "Interferometric Astrometry of Proxima Centauri and Barnard's Star Using *Hubble Space Telescope* Fine Guidance Sensor 3: Detection Limits for Substellar Companions," *Astronomical Journal* 118 (1999), 1086-1100.

12. (*Chapter 2*)
See for example, P. M. W. Kalbera and K. Jürgen, "The HI Distribution of the Milky Way," *Annual Review of Astronomy and Astrophysics* 47, 1 (2009), 27-61, and references therein.

13. (*Chapter 2*)
A study in which expected limits on the orbits and masses of exomoons were estimated: C. Weidner and K. Horne, "Limits on the Orbits and Masses of Moons around Currently-known Transiting Exoplanets," *Astronomy and Astrophysics* 521 (2010), A76, doi: 10.1051/0004-6361/201014955. See also, C. S. Gaudi, "Microlensing by Exoplanets," in *Exoplanets* (Arizona University Press, 2010), 101, eds. Sara Seager and Renée Dotson, and references therein.

14. (*Chapter 2*)
The free-floating objects were found using the "microlensing method." T. Sumi, et al., "Unbound or Distant Planetary Mass Population Detected by Gravitational Microlensing," *Nature* 473 (2011), 349-352.

Chapter 3

1. (*Chapter 3*)
See, for example, various discussions in R. S. Ball, *Great Astronomers* (Pitman, 1907).

2. (*Chapter 3*)
See, for example, various discussions on the spectroscopy of stars in R. O. Gray, C. J. Corbally, and A. J. Burgasser, *Stellar Spectral Classification* (Princeton University Press, 2009). In order of decreasing temperature, stars are classified by the letters O, B, A, F, G, K, M (our Sun is a G star). Note that the first distance measured to a star other than our Sun was achieved in 1838, see: F. W. Bessel, "On the Parallax of 61 Cygni," *Monthly Notices of the Royal Astronomical Society* 4 (1838), 152-161. A milestone in the understanding of the nuclear physics in stars was reached in the late 1950s, marked by the seminal paper: E. M. Burbidge, G. R. Burbidge, W. A. Fowler, and F. A. Hoyle, "Synthesis of the Elements," *Reviews of Modern Physics* 29, 4 (1957), 547-650.

3. (*Chapter 3*)
For further reading, see for example: G. Shaviv, *The Life of Stars: The Controversial Inception and Emergence of the Theory of Stellar Structure*, (Springer, 2009); C. Iliadis, *Nuclear Physics of Stars* (Wiley-VCH, 2007).

4. (*Chapter 3*)

A. Wolszczan and D. A. Frail, "A Planetary System around the Millisecond Pulsar PSR1257+12," *Nature* 335 (1992), 145.

5. (*Chapter 3*)

G. H. Walker, "Doppler Searches for Planetary Systems," *Astrophysics and Space Science* 223, 2 (1995), 103-107.

6. (*Chapter 3*)

M. Mayor and D. Queloz, "A Jupiter-mass Companion to a Solar-type Star," *Nature* 378 (1995), 355-359.

7. (*Chapter 3*)

See, for example, G. Anglada-Escudé, et al., "Strong Constraints on the Putative Planet Candidate Around VB10 Using Doppler Spectroscopy," *Astrophysical Journal Letters* 711 (2010), L24-L29; J. L. Bean, et al., "The Proposed Giant Planet Obriting VB10 Does Not Exist," *Astrophysical Journal Letters* 711 (2010), L19-L23.

8. (*Chapter 3*)

The most sensitive astrometry observations so far have been done with the *Hubble Space Telescope* and the *Hipparcos satellite*. See, for example, A. Quirrenbach, "Astrometric Detection and Characterization of Exoplanets," in *Exoplanets* (Arizona University Press, 2010), 166, eds. Sara Seager and Renée Dotson.

9. (*Chapter 3*)

For a sample of randomly oriented star-planet systems, the expected probability of detecting a transit can be shown to be approximately equal to the ratio of the stellar radius and the semi-major axis of the orbit. However, the detailed calculations involve more parameters and complexity. A good quantitative overview of analyzing and interpreting exoplanet transit data can be found in: J. N. Winn, "Exoplanet Transits and Occultations," in *Exoplanets* (Arizona University Press, 2010), 55-77, eds. Sara Seager and Renée Dotson. See also, C. A. Haswell, *Transiting Exoplanets*, (Cambridge University Press, 2010).

10. (*Chapter 3*)
G. W. Henry, G. W. Marcy, R. P. Butler, and S. S. Vogt, "A Transiting '51 Peg-like' Planet" *Astrophysical Journal* 529 (2000), L41-L44; D. Charbonneau, T. M. Brown, D. W. Latham, and M. Mayor, "Detection of Planetary Transits Across a Sun-like Star," *Astrophysical Journal* 529 (2000), L45-L48.

11. (*Chapter 3*)
By the 8th of October 2011, 25 exoplanets in 22 alien solar systems had been directly imaged. Check the *Extrasolar Planets Encyclopedia* for updated information. The first exoplanet to be directly imaged is known as Fomalhaut b, observed by the *Hubble Space Telescope*. See P. Kalas, et al., "Optical Images of an Extrasolar Planet 25 Light Years from Earth," *Science* 322 (2008), 1345-1348. Some exoplanets have been imaged at infrared wavelengths as opposed to optical (visible) wavelengths. For a review on exoplanet imaging, see W. A. Traub and B. R. Oppenheimer, "Direct Imaging of Exoplanets," in *Exoplanets* (Arizona University Press, 2010), 111-156, eds. Sara Seager and Renée Dotson.

12. (*Chapter 3*)
W. A. Traub and B. R. Oppenheimer, "Direct Imaging of Exoplanets," in *Exoplanets* (Arizona University Press, 2010), 111-156, eds. Sara Seager and Renée Dotson.

13. (*Chapter 3*)

D. W. Letham, et al., "The Unseen Companion of HD 114762: A Probable Brown Dwarf," *Nature* 339 (1989), 38-40.

Chapter 4

1. (*Chapter 4*)

See, for example, R. O. Gray, C. J. Corbally, and A. J. Burgasser, *Stellar Spectral Classification* (Princeton University Press, 2009).

2. (*Chapter 4*)

F. Selsis, et al., "Could We Identify Hot Ocean-Planets with *CoRoT*, *Kepler* and Doppler Velocimetry?" *Icarus*, 191, 2 (2007), 453-468.

3. (*Chapter 4*)

The detection of atmospheric escape from HD 209458b was first reported in: A. Vidal-Madjar, et al., "An Extended Upper Atmosphere around the Extrasolar Planet HD 209458b," *Nature* 422 (2003), 143-146. A refutation of the atmospheric escape scenario is given in, L. Ben-Jaffel, G. Ballester, J. T. Clarke, and F. Vincent, "Exoplanet HD 209458b: Inflated Hydrogen Atmosphere but No Sign of Evaporation," *Astrophysical Journal* 671 (2007), L61-L64. A rebuttal appeared in: A. Vidal-Madjar, et al., "Exoplanet HD 209458b (*Osiris*): Evaporation Strengthened," *Astrophysical Journal* 676 (2008), L57-L60. A general discussion can be found in: J. J. Fortney, I. Baraffe, and B. Militzer, "Giant Planet Interior Structure and Thermal Evolution," *Exoplanets* (Arizona University Press, 2010), 411, eds. Sara Seager and Renée Dotson. Some theoretical aspects of the unusual atmosphere of HD 209458b can be found in: A. Claret, "Does the HD 209458 Planetary System

Pose a Challenge to the Stellar Atmosphere Models?" *Astronomy and Astrophysics* 506 (2009), 1335-1340; R. A. Murray-Clay, E. I. Chiang, and N. Murray, "Atmospheric Escape from Hot Jupiters," *Astrophysical Journal* 693 (2009), 23-42; J. I. Moses, et al., "Disequilibrium Carbon, Oxygen, and Nitrogen Chemistry in the Atmospheres of HD 189733b and HD 209458b," *Astrophysical Journal* 737 (2011), 15-54.

4. (*Chapter 4*)

The quotation is from J. J. Fortney, I. Baraffe, and B. Militzer, "Giant Planet Interior Structure and Thermal Evolution," in *Exoplanets* (Arizona University Press, 2010), 409, eds. Sara Seager and Renée Dotson. A review of the proposed mechanisms to account for the radius anomaly in close-in hot Jupiters can be found in the same paper. Some studies pointing out the tidal instability and/or destruction of close-in planets: B. Jackson, R. Greenberg, and R. Barnes, "Tidal Evolution of Close-in Extrasolar Planets," *Astrophysical Journal* 678 (2008), 1396-1406; B. Jackson, R. Barnes, and R. Greenberg, "Observational Evidence for the Tidal Destruction of Extrasolar Planets," *Astrophysical Journal* 698 (2009), 1357-1366; B. Levrard, C. Winisdoerffer, and G. Chabrier, "Falling Transiting Extrasolar Giant Planets," *Astrophysical Journal Letters* 692 (2009), L9-L13.

5. (*Chapter 4*)

See: S. Udry, M. Mayor, and N. C. Santos, "Statistical Properties of Exoplanets I. The Period Distribution: Constraints on the Migration Scenario," *Astronomy and Astrophysics* 407 (2003), 369-376; H. R. A. Jones, "An Exoplanet in Orbit Around τGruis," *Monthly Notices of the Astronomical Society* 341 (2003), 948-952.

6. (*Chapter 4*)

See, for example: D. N. C. Lin, P. Bodenheimer, and D. C. Richardson, "Orbital Migration of the Planetary Companion of 51 Pegasi to its Present Location," *Nature* 380 (1996), 606-607; M. J. Kuchner and M. Lecar, "Halting Planet Migration in the Evacuated Centers of Protoplanetary Disks," *The Astrophysical Journal* 574 (2002), L87-89; M. Nagasawa and T. Bessho, "Formation of Hot Planets by a Combination of Planet Scattering, Tidal Circularization, and the Kozai Mechanism," *The Astrophysical Journal* 678 (2008), 498-508; T. A. Davis and P. J. Wheatley, "Evidence for a Lost Population of Close-in Exoplanets," *Monthly Notices of the Astronomical Society* 396 (2009), 1012-1017; B. Jackson, R. Barnes, and R. Greenberg, "Observational Evidence for Tidal Destruction of Exoplanets," *The Astrophysical Journal* 698 (2009), 1357-1366.

7. (*Chapter 4*)

P. Beníez-Llambay, F. Masset, and C. Beaugé, "The Mass-Period Distribution of Close-in Exoplanets," *Astronomy and Astrophysics* 528 (2011), doi:10.1051/004-6361/20105774.

8. (*Chapter 4*)

See A. Cumming, "Statistical Properties of Exoplanets," in *Exoplanets* (Arizona University Press, 2010), 203-205, eds. Sara Seager and Renée Dotson, and references therein.

9. (*Chapter 4*)

Spin-orbit misalignment cannot be directly measured and must be inferred from statistically "deprojecting" indirect measurements on the sky. A recent study is described in A. H. M. J. Triaud et al., "Spin-orbit Angle Measurements for Six Southern Transiting Planets. New insights into the Dynamical Origins of Hot Jupiters,"

Astronomy and Astrophysics 524 (2010), A25, doi: 10.1051/0004-6361/201014525. The authors conclude that there are 26 transiting exoplanets that have sky-projected obliquities, and 8 of these show significant spin-orbit misalignments. Of the 8, 5 show apparent retrograde motion. From a statistical simulation analysis the authors then conclude that 85% of hot Jupiters actually have spin-orbit misalignment. In S. Naoz, et al., "Hot Jupiters from Secular Planet-Planet Interactions," *Nature* 473 (2011), 187-189, the authors show calculations demonstrating that long-term interactions of close-in hot Jupiters with more distant massive planets (or brown dwarfs) can explain spin-orbit misalignment as well as retrograde motion. However, the model makes the very strong prediction that in *every* system that harbors a hot Jupiter, there *has to be* a more distant massive object present in that system. A single experimental verification of this prediction has yet to be made. As for exoplanets with a high eccentricity, see, for example, P. Arriagada, et al., "Five Long-period Extrasolar Planets in Eccentric Orbits from the Magellan Planet Search Program," *Astrophysical Journal* 711 (2010), 1229-1235, and references therein. The paper reiterates that the reason for exoplanets having an eccentricity distribution that is skewed to a nonzero value (a median value of 0.24 is quoted) is still being debated. Planet-planet scattering is invoked as a possible explanation, but as with spin-orbit misalignment, a second object responsible for the interaction has never been found in any system.

10. (*Chapter 4*)
See, for example, J. E. Chambers, "Making More Terrestrial Planets," *Icarus* 152, 2 (2001), 205-224, and references therein.

11. (*Chapter 4*)

A. W. Howard, et al., "The Occurrence and Mass Distribution of Close-in Super-Earths, Neptunes, and Jupiters," *Science* 330 (2010), 653-655. See also, A. W. Howard, et al., "Planet Occurrence within 0.25 AU of Solar-type Stars from *Kepler*," (2011), http://arxiv.org/pdf/1103.2541v1.

12. (*Chapter 4*)

J. E. Chambers, "Terrestrial Planet Formation," in *Exoplanets* (Arizona University Press, 2010), 297-317, eds. Sara Seager and Renée Dotson.

13. (*Chapter 4*)

G. Laughlin and P. Bodenheimer, "Nonaxisymmetric Evolution in Protostellar Disks," *Astrophysical Journal* 436 (1994), 330-354.

14. (*Chapter 4*)

G. D'Angelo, R. H. Durisen, and J. J. Lissauer, "Giant Planet Formation," in *Exoplanets* (Arizona University Press, 2010), 334-341, eds. Sara Seager and Renée Dotson.

15. (*Chapter 4*)

See, for example, E. I. Vorobyov, "Embedded Protostellar Disks around (Sub-)Solar Stars. II. Disk Masses, Sizes, Densities, Temperatures and the Planet Formation Perspective," *Astrophysical Journal* 729 (2011) http://arxiv.org/pdf/1101.3090v1, and references therein.

16. (*Chapter 4*)

See S. Nayakshin, "Formation of Planets by Tidal Downsizing of Giant Planet Embryos," *Monthly Notices of the Royal Astronomical Society* 408 (2010), L36-L40; S. Nayakshin, "Formation of

Terrestrial Planet Cores inside Giant Planet Embryos," *Monthly Notices of the Royal Astronomical Society* 413 (2011), 1462-1478, and references therein.

17. (*Chapter 4*)
D. N. C. Lin, P. Bodenheimer, and D. C. Richardson, "Orbital Migration of the Planetary Companion of 51 Pegasi to its Present Location," *Nature* 380 (1996), 606-607.

18. (*Chapter 4*)
Spin-orbit misalignment cannot be directly measured and must be inferred from statistically "deprojecting" indirect measurements on the sky. A recent study is described in A. H. M. J. Triaud, et al. "Spin-orbit Angle Measurements for Six Southern Transiting Planets. New Insights into the Dynamical Origins of hot Jupiters," *Astronomy and Astrophysics* 524 (2010), A25, doi: 10.1051/0004-6361/201014525. The authors conclude that there are 26 transiting exoplanets that have sky-projected obliquities, and 8 of these show significant spin-orbit misalignments. Of the 8, 5 show apparent retrograde motion. From a statistical simulation analysis the authors than conclude that 85% of hot Jupiters actually have spin-orbit misalignment. In S. Naoz, et al., "Hot Jupiters from secular planet-planet interactions," *Nature* 473 (2011), 187-189, the authors show calculations demonstrating that long-term interactions of close-in hot Jupiters with more distant massive planets (or brown dwarfs) can explain spin-orbit misalignment as well as retrograde motion. However, the model makes the very strong prediction that in *every* system that harbors a hot Jupiter, there *has to be* a more distant massive object present in that system. A single experimental verification of this prediction has yet to be made. As for exoplanets with a high eccentricity, see, for example, P. Ar-

riagada, et al., "Five Long-period Extrasolar Planets in Eccentric Orbits from the Magellan Planet Search Program," *Astrophysical Journal* 711 (2010), 1229-1235, and references therein. The paper reiterates that the reason for exoplanets having an eccentricity distribution that is skewed to a nonzero value (a median value of 0.24 is quoted), is still being debated. Planet-planet scattering is invoked as a possible explanation, but as with spin-orbit misalignment, a second object responsible for the interaction has never been found in any system.

19. (*Chapter 4*)
The quote is from the Triaud, et al. paper in the preceding note.

20. (*Chapter 4*)
See B. Levrard, C. Winisdoerffer, and G. Chabrier, "Falling Transiting Extrasolar Giant Planets," *Astrophysical Journal Letters* 692 (2009), L9-L13. For the tendency of the closest-in hot Jupiters to be associated with the youngest host stars, see B. Jackson, R. Barnes, and R. Greenberg, "Observational Evidence for the Tidal Destruction of Extrasolar Planets," *Astrophysical Journal* 698 (2009), 1357-1366.

21. (*Chapter 4*)
See, for example, A. Cumming, "Statistical Properties of Exoplanets," in *Exoplanets* (Arizona University Press, 2010), 205-208, eds. Sara Seager and Renée Dotson, and references therein.

22. (*Chapter 4*)
G. M. Kennedy and S. J. Keynon, "Planet Formation around Stars of Various Masses: Hot Super-Earths," *The Astrophysical Journal* 682 (2008), 1264-1276.

23. (*Chapter 4*)

P. R. Butler, et al., "A Long-Period Jupiter-Mass Planet Orbiting the Nearby M Dwarf GJ 849," *Publication of the Astronomical Society of the Pacific* 118 (2006), 1685-1689; A. Cumming, et al., "The Keck Planet Search: Detectability and the Minimum Mass and Orbital Period Distribution of Extrasolar Planets," *Publication of the Astronomical Society of the Pacific* 120 (2008), 531-554.

24. (*Chapter 4*)

J. A. Johnson, et al., "Retired A Stars and Their Companions: Exoplanets Orbiting Three Intermediate-Mass Subgiants," *Astrophysical Journal* 665 (2007), 785-793; J. A. Johnson, et al., "Retired A Stars and Their Companions. II. Jovian Planets Orbiting κCrB and HD 167042," *Astrophysical Journal* 675 (2008), 784-789. See also, A. Cumming, "Statistical Distribution of Exoplanets," in *Exoplanets* (Arizona University Press, 2010), 206-208, and references therein.

25. (*Chapter 4*)

D. A. Fischer and J. Velenti, "The Planet-Metallicity Correlation," *The Astrophysical Journal* 622 (2005), 1102-1117; A. Cumming, "Statistical Properties of Exoplanets," in *Exoplanets* (Arizona University Press, 2010), 205-206, eds. Sara Seager and Renée Dotson, and references therein.

26. (*Chapter 4*)

I. Ramírez, et al., "A Possible Signature of Terrestrial Planet Formation in the Chemical Composition of Solar Analogs," *Astronomy and Astrophysics* 521 (2010), 33-44.

27. (*Chapter 4*)

See, for example, W. Kley, M. H. Lee, N. Murray, and S. J. Peale,

"Modeling the Resonant Planetary System GJ 876," *Astronomy and Astrophysics* 437 (2005), 727-742.

28. (*Chapter 4*)
W. J. Borucki, et al., "Characteristics of *Kepler* Planetary Candidates Based on the First Data Set," *Astrophysical Journal* 728 (2011), 117-137; W. J. Borucki, et al., "Characteristics of Planetary Candidates Observed by *Kepler*, II: Analysis of the First Four Months of Data," *Astrophysical Journal* 736 (2011), doi: 10.1088/004-637X/736/1/19.

Chapter 5

1. (*Chapter 5*)
Amino acids are composed of a class of molecules that consists of a common group of atoms (with a particular spatial configuration) plus a component ("side chain") that is different for different amino acids. The latter gives an amino acid unique properties compared to other amino acids. Amino acids are the building blocks of proteins, and there are 20 different amino acids to make the proteins required for life. Proteins can be very large molecules, consisting of hundreds to thousands of amino acids. The particular combinations of amino acids, and their relative positions in a complex molecule give rise to different proteins with unique properties.

2. (*Chapter 5*)
For example, see discussion in C. A. Scharf, *Extrasolar Planets and Astrobiology* (University Science Books, 2009), 443-452.

3. (*Chapter 5*)

See, for example, F. Dyson, *Origins of Life* (Cambridge University Press, 2004), in which both the Oparin and Eigen scenarios are described, as well as a third by Cairns-Smith. The latter involves information contained in the patterns of ions in a clay substrate that may be transferred to chemical species dissolved in the water. However, there is no driver for replication and metabolism to develop, and statistical fluctuations are again the only mechanism. F. Dyson also discusses his own scenario, which is a hybrid of the Oparin and Eigen scenarios, advocating a dual origin. However, this does not solve the problem of the origin of life either.

4. (*Chapter 5*)

F. Dyson, *Origins of Life* (Cambridge University Press, 2004), 38-44.

5. (*Chapter 5*)

For example, see G. Fried and G. Hademenos, *Schaum's Outline of Biology* (McGraw Hill Professional, 2009), 370-375.

6. (*Chapter 5*)

Eigen scenario computer simulations: U. Niesert, D. Harnasch, and C. Bresch, "Origin of Life between Scylla and Charybdis," *Journal of Molecular Evolution* 17 (1981), 348-353. See also discussion in F. Dyson, *Origins of Life* (Cambridge University Press, 2004), 41-42.

7. (*Chapter 5*)

G. Fried and G. Hademenos, *Schaum's Outline of Biology* (McGraw Hill Professional, 2009), 371.

8. (*Chapter 5*)

F. Dyson, *Origins of Life* (Cambridge University Press, 2004), 44.

9. (*Chapter 5*)

See, for example, M. S. Longair, *Theoretical Concepts in Physics* (University of Cambridge Press, 1984).

10. (*Chapter 5*)

For example, see F. Dyson, *Origins of Life* (Cambridge University Press, 2004), 8.

11. (*Chapter 5*)

W. Gilbert, "The RNA World," *Nature* 319 (1986), 618.

12. (*Chapter 5*)

S. J. Mojzsis, et al., "Evidence for Life on Earth Before 3800 Million Years Ago," *Nature* 384 (1996), 55-59.

13. (*Chapter 5*)

Papers describing the original experiments that attempted to make amino acids under primordial conditions: S. Miller, "Production of Amino Acids Under Possible Primitive Earth Conditions," *Science* 117 (1953), 3046, 528-529; S. Miller and H. C. Urey, "Organic Compound Synthesis on the Primitive Earth," *Science* 130 (1959), 3370, 245-251.

14. (*Chapter 5*)

See, for example, C. Nouvian, *The Deep: The Extraordinary Creatures of the Abyss*, (University of Chicago Press, 2007).

15. (*Chapter 5*)

See, for example, M. J. Russell, R. M. Daniel, A. J. Hall, and J. Sherringham, "A Hydrothermally Precipitated Catalytic Iron Sulphide Membrane as a First Step Toward Life," *Journal of Molecular Evolution* 39 (1994), 231-243.

16. (*Chapter 5*)
K. O. Stetter, "History of Discovery of the First Hyperthermophiles," *Extremophiles* 10, 5 (2006), 357-362.

17. (*Chapter 5*)
D. N. Thomas and G. S. Dieckman, "Antarctic Sea Ice - a Habitat for Extremophiles," *Science* 295, 5555 (2002), 641-644.

18. (*Chapter 5*)
G. Borgonie, et al., "Nematoda from the Terrestrial Deep Subsurface of South Africa," *Nature* 474 (2011), 79-82.

19. (*Chapter 5*)
See, for example, discussion in C. A. Scharf, *Extrasolar Planets and Astrobiology*, (University Science Books, 2009), 345-347. You may have heard that an excess of left-handed amino acids has been found in meteorites. One of the articles about this appeared in the *Astrobiology Magazine* in 2011 (http://www.astrobio.net/pressrelease/3748/) and was entitled, "Meteorites May Answer Life's Chirality Question" (other articles appeared in different places as usual, based on the same information, and the story in such cases is usually simply replicated many times over). The article is very strange and is a fine example of extreme handwaving. Even before reading the article, you should realize that the meteorites alone cannot answer the question because the question is simply deferred. How was the bias introduced in the meteorites? The article admits that the mechanism is not known, and postulates that primordial water could be responsible for the asymmetry. However, no actual mechanism is

proposed. Then it gets worse. The article admits that even if water could do it, it can only amplify an already existing asymmetry, and could not create one. Having admitted that, the article then postulates that radiation could have produced the asymmetry, but no mechanism is proposed. There is no reason why radiation should introduce an asymmetry. It has never been demonstrated in the laboratory, or theoretically (the article does not cite any study). The article then ends with a pretty graphic of an artist's impression of a neutron star, which has nothing to do with amino acids, let alone left-handed ones. The only connection is that the neutron star produces radiation, but a large variety of other objects in the Universe produce radiation too. The *Wikipedia* article on amino-acid chirality is even stranger. It states that "Most scientists believe that Earth life's 'choice' of chirality was purely random." This is nonsense because random processes produce equal numbers of left-handed and right-handed amino acids. That is the point! In order to introduce a bias, a nonrandom process is explicitly required, and that process has not been identified. So, scientists who believe in the "random choice" (and such scientists *do* exist) actually believe in nothing, no explanation at all, because the mechanism producing a choice in the first place is not known.

20. (*Chapter 5*)
F. Wolfe-Simon, et al., "A Bacterium That Can Grow by Using Arsenic Instead of Phosphorus," *Science Express* (2010), 1-4, doi: 10.1126/science.1197528.

21. (*Chapter 5*)
A. E. Saal, et al., "Volatile Content of Lunar Volcanic Glasses and the Presence of Water in the Moon's Interior," *Nature* 454 (2008), 192-195.

22. (*Chapter 5*)
R. N. Clark, "Detection of Adsorbed Water and Hydroxyl on the Moon," *Science* 326 (2009), 562-564.

23. (*Chapter 5*)
E. H. Hauri, et al., "High Pre-Eruptive Water Contents Preserved in Lunar Melt Inclusions," *Science Express* (2011), 1-4, doi: 10.1126/science.1204626.

24. (*Chapter 5*)
R. M. Canup, "Dynamics of Lunar Formation," *Annual Review of Astronomy and Astrophysics* 42 (2004), 441-475, and references therein.

25. (*Chapter 5*)
S. Nayakshin, "Rotation of the Solar System Planets and the Origin of the Moon in the Context of the Tidal Downsizing Hypothesis," *Monthly Notices of the Astronomical Society* 410 (2011), L1-L5.

26. (*Chapter 5*)
C. R. Javoy, "Where Do the Oceans Come From?" *Geoscience* 337 (2005), 139-158.

27. (*Chapter 5*)
See, for example, D. Hutsemékers, J. Manfroid, E. Jehin, and C. Arpigny, "New Constraints on the Delivery of Cometary Water and Nitrogen to Earth from the 15N/14N Isotopic Ratio," *Icarus* 204 (2009), 346-348; H. Genda and M. Ikoma, "Origin of the Ocean on the Earth: Early Evolution of Water D/H in a Hydrogen-rich Atmosphere," *Icarus* 194 (2008), 42-52.

28. (*Chapter 5*)

H. Hsieh, "A Frosty Finding," *Nature* 464 (2010), 1286-1287.

29. (*Chapter 5*)

G. W. Swift and D. A. Geller, "Thermoacoustics and Thermal Dissociation of Water," *Journal of the Acoustical Society of America* 122, 5 (2007), 1, doi: 10.1121/1.2942772.

30. (*Chapter 5*)

See, for example, C. A. Scharf, *Extrasolar Planets and Astrobiology*, (University Science Books, 2009), 280-295.

31. (*Chapter 5*)

D. Charbonneau, T. M. Brown, R. W. Noyles, and R. L. Gilliland, "Detection of an Extrasolar Planet Atmosphere," *Astrophysical Journal* 568 (2002), 377-384; S. Seager and D. Deming, "Exoplanet Atmospheres," *Annual Reviews of Astronomy and Astrophysics* 48 (2010), 631-672, and references therein.

32. (*Chapter 5*)

Here, polarization refers to the directional plane of the electric and magnetic fields in rays of light. Normally, light consists of rays with randomly oriented planes of the electromagnetic field and such light is unpolarized. Scattering by electrons can selectively pick out preferred orientations of the electromagnetic field and thereby polarize light. In the context of detecting polarized light from scattering in an exoplanet atmosphere, see, for example, S. V. Berdyugina, A. V. Berdyugin, D. M. Fluri, and V. Piirola, "Polarized Reflected Light from the Exoplanet HD 189733b: First Multicolor Observations and Confirmation of Detection," *Astrophysical Journal* 728 (2011), L6-L10, http://arxiv.org/pdf/1101.0059v1, and references therein. The preceding work answers an earlier criticism in the paper by S. J. Wiktorowicz, "Nondetection of Polarized, Scattered Light from the HD 189733b Hot Jupiter," *Astrophysical Journal* (2009), 1116-1124.

33. (*Chapter 5*)
S. Seager and D. Deming, "Exoplanet Atmospheres," *Annual Reviews of Astronomy and Astrophysics* 48 (2010), 631-672, and references therein.

34. (*Chapter 5*)
J. L. Bean, E. M.-R. Kempton, and D. Homeier, "A Ground-based Transmission Spectrum of the Super-Earth Exoplanet GJ 1214b," *Nature* 468 (2010), 669-672.

35. (*Chapter 5*)
See, for example, S. Seager and D. Deming, "Exoplanet Atmospheres," *Annual Reviews of Astronomy and Astrophysics* 48 (2010), 631-672; C. A. Scharf, *Extrasolar Planets and Astrobiology*, (University Science Books, 2009), 271-280.

36. (*Chapter 5*)
J. Marotzke and M. Botzet, "Present-day and Ice-covered Equilibrium States in a Comprehensive Climate Model," *Geophysical Research Letters* 34, 16 (2007), L16704, doi: 10.1029/2006GL028880; see also D. Spiegel, K. Menou, and C. A. Scharf, "Habitable Climates," *Astrophysical Journal* 681 (2008), 1609-1623.

37. (*Chapter 5*)
C. Sagan and G. Mullen, "Earth and Mars: Evolution of Atmospheres and Surface Temperatures," *Science* 177, 4043 (1972), 52-

56.

38. (*Chapter 5*)
The following paper claimed that the "faint young Sun paradox" had been resolved, on the basis of changing the reflectivity (albedo) of clouds and the Earth's surface: M. T. Rosing, D. K. Bird, and N. H. Sleep, "No Climate Paradox Under the Faint Early Sun," *Nature* 464 (2010), 744-747. That result has been refuted on the basis that even with the most favorable assumptions that would raise the Earth's temperature, the increased heat energy retained by Earth falls short by a factor of two (compared to that required to solve the problem). In other words, the refutation asserts that the cloud-changing assumptions made in the Rosing, et al. (2010) paper were not justified, and are not achievable. These refutations were published in: C. Goldblatt and K. J. Zahnle, "Faint Young Sun Paradox Remains," (2011) http://arxiv.org/abs/1105.5425, doi: 10.1038/nature09961.

39. (*Chapter 5*)
J. F. Kasting, D. P. Whitmire, and R. T. Reynolds, "Habitable Zones around Main Sequence Stars," *Icarus* 101, 1 (1993), 108-128. A recent review can be found in, H. Lammer, et al., "What Makes a Habitable Planet?" *Annual Reviews of Astronomy and Astrophysics* 17 (2009), 181-249.

40. (*Chapter 5*)
A good recent discussion of the complex properties of water can be found in, C. A. Scharf, *Extrasolar Planets and Astrobiology*, (University Science Books, 2009), 366-371.

41. (*Chapter 5*)
See E. J. Carpenter, S. Lin, and D. G. Capone, "Bacterial Activity

in South Pole Snow," *Applied and Environmental Microbiology* 66 (2000), 4514-4517; K. Kashefi and D. R. Lovley, "Extending the Upper Temperature Limit for Life," *Science* 301, 5635 (2003), 934.

42. (*Chapter 5*)
D. S. Spiegel, K. Menou, and C. A. Scharf, "Habitable Zones," *Astrophysical Journal* 681 (2008), 1609-1623. This paper has a good general discussion of the problems with the conventional definition of the habitable zone, and the standard criteria for habitability. The paper points out that "average" planet properties are insufficient to determine habitability. At the very least, seasonal effects should be taken into account (which, for some planets may be very extreme). The authors also emphasize that, as far as observations go, the fractional habitability (which has both spatial and temporal dependences), must be large enough to create observable biosignatures.

43. (*Chapter 5*)
D. S. Spiegel, K. Menou, and C. A. Scharf, "Habitable Climates: The Influence of Obliquity," *Astrophysical Journal* 691 (2009), 596-610. Simulations in this paper show that for oblique orbits there are large seasonal variations, and equilibrium models may not be appropriate. However such planets may support local regions that are habitable over a larger range of orbital radii than Earth.

44. (*Chapter 5*)
W. von Bloh, M. Cuntz, S. Franck, and C. Bounama, "Habitability of the Goldilocks Planet Gliese 581g: Results from Geodynamic Models," *Astronomy and Astrophysics* 528 (2011), A133, doi: 10.1051/004-6361/201116534.

45. (*Chapter 5*)

F. Selsis, et al., "Habitable Planets around the Star Gliese 581?" *Astronomy and Astrophysics* 476 (2007), 1373-1387.

46. (*Chapter 5*)

W. von Bloh, M. Cuntz, S. Franck, and C. Bounama, "Habitability of the Goldilocks Planet Gliese 581g: Results from Geodynamic Models," *Astronomy and Astrophysics* 528 (2011), A133, doi: 10.1051/004-6361/201116534; P. von Paris, et al., "Atmospheric Studies of Habitability in the Gliese 581 System," *Astronomy and Astrophysics* (2011), http://arxiv.org/pdf/1104.3756v1, and references therein; L. Kaltenegger, A. Segura, and S. Mohanty, "Model Spectra of the First Potentially Habitable Super-Earth," *Astrophysical Journal* 733, 1 (2011), doi: 10.1088/0004-637X/733/1/35, and references therein.

47. (*Chapter 5*)

F. Selsis, et al., "Habitable Planets Around the Star Gliese 581?" *Astronomy and Astrophysics* 476 (2007), 1373-1387; W. von Bloh, M. Cuntz, S. Franck, and C. Bounama, "Habitability of the Goldilocks Planet Gliese 581g: Results from Geodynamic Models," *Astronomy and Astrophysics* 528 (2011), A133, doi: 10.1051/004-6361/201116534; Y. Hu and F. Ding, "Radiative Constraints on the Habitability of Exoplanets Gliese 581c and Gliese 581d," *Astrophysical Journal* 526 (2011), doi: 10.1051/0004-6261/201014880.

48. (*Chapter 5*)

W. J. Borucki, et al., "Characteristics of *Kepler* Planetary Candidates Based on the First Data Set," *Astrophysical Journal* 728 (2011), 117-137; W. J. Borucki, et al., "Characteristics of Planetary Candidates Observed by *Kepler*, II: Analysis of the First

Four Months of Data," *Astrophysical Journal* 736 (2011), doi: 10.1088/004-637X/736/1/19.

49. (*Chapter 5*)
N. Prantzos, "On the Galactic Habitable Zone," *Space Science Reviews* 135 (2008), 313-322, doi: 10.1007/s11214-007-9236-0. The concept of the "Galactic Habitable Zone" was proposed by G. Gonzalez, D. Brownlee, and P. Ward, "The Galactic Habitable Zone: Galactic Chemical Evolution," *Icarus* (2001), 152, 185-200.

50. (*Chapter 5*)
See, for example, P. D. Ward and D. Brownlee, *Rare Earth: Why Complex Life is Uncommon in the Universe*, (Springer, 2000).

51. (*Chapter 5*)
For example, D. Darling, *Life Everywhere: The Maverick Science of Astrobiology*, (Basic Books, 2002).

52. (*Chapter 5*)
For example,
http://seti.org and http://setisociety.org.

53. (*Chapter 5*)
J. Catanzarite and M. Shao, "The Occurrence Rate of Earth Analog Planets Orbiting Sunlike Stars," (2011), http://arxiv.org/pdf/1103.1443v2. Note that the estimates of the occurrence of Earth analogs here refer specifically to planets orbiting stars of type F, G, or K. (In order of decreasing temperature, stars are classified using the letters O, B, A, F, G, K, M; our Sun is a G star.) The paper notes a particular caveat for uncertainties in the analysis for very hot stars. The conservative estimate of the occurrence of Earth analogs was made using the habit-

able zone assumptions of J. F. Kasting, D. P. Whitmore, and R. T. Reynolds, "Habitable Zones around Main Sequence Stars," *Icarus* 101, 1 (1993), 108-128. The more optimistic estimate of the occurrence of Earth analogs was made using the assumptions in J. Lunine, *Exoplanet Task Force Report*, (2008).

Appendix A

1. (*Appendix A*)
See, for example, T. Sumi, et al., "Unbound or Distant Planetary Mass Population Detected by Gravitational Microlensing," *Nature* 473 (2011), 349-352.

2. (*Appendix A*)
W. B. Hubbard, A. Burrows, and J. I. Lunine, "Theory of Giant Planets," *Annual Reviews of Astronomy and Astrophysics*, 40 (2002), 103-136. See also a note on the IAU website at http://www.iau.org/public_press/news/release/iau0601/ q_answers/.

3. (*Appendix A*)
See: http://web.gps.caltech.edu/~mbrown/dwarfplanets/.

4. (*Appendix A*)
W. B. Hubbard, A. Burrows, and J. I. Lunine, "Theory of Giant Planets," *Annual Reviews of Astronomy and Astrophysics*, 40 (2002), 103-136.

Index

About the Author

Dr. Tahir Yaqoob is an astrophysicist and educator with over 25 years of experience. He obtained a B.A. and an M.A. in physics from the University of Oxford, England, and a Ph.D. in astrophysics from the University of Leicester, England. He has tutored and mentored students across the entire academic range, from students at elementary school to those in Ph.D. programs. He has also trained postgraduate students and postdoctoral researchers to become established scientists and professors in physics and astrophysics. Dr. Yaqoob has published over 120 research papers in peer-reviewed international journals and is the author of the book *What Can I Do to Help My Child with Math When I Don't Know Any Myself?* He works on NASA-funded astrophysics research projects and he is also a member of the editorial board of the international peer-reviewed journal *ISRN Astronomy and Astrophysics*. Dr. Yaqoob currently lives in Baltimore, USA, with his wife, Rehana, and two children, Humza and Aazam.

CPSIA information can be obtained at www.ICGtesting.com
Printed in the USA
BVOW08s0814180715

408917BV00002B/45/P